Birkhäuser

Alex Iosevich • Elijah Liflyand

Decay of the Fourier Transform

Analytic and Geometric Aspects

Birkhäuser

Alex Iosevich
Department of Mathematics
University of Rochester
Rochester, NY, USA

Elijah Liflyand
Department of Mathematics
Bar-Ilan University
Ramat-Gan, Israel

ISBN 978-3-0348-0624-4 ISBN 978-3-0348-0625-1 (eBook)
DOI 10.1007/978-3-0348-0625-1
Springer Basel Heidelberg New York Dordrecht London

Library of Congress Control Number: 2014952035

Mathematics Subject Classification (2010): 42A, 42B

Printed on acid-free paper

Springer Basel is part of Springer Science+Business Media (www.birkhauser-science.com)

TO OUR FAMILIES

למשפחותינו בהוקרה

Contents

Foreword

In 1920, in volume **27**, page 175 of the American Mathematical Monthly the following sonnet, written by Sir William Rowan Hamilton and dedicated to Fourier's heritage and personality was published.

Fourier with solemn and profound delight,
Joy born of awe, but kindling momently
To an intense and thrilling ecstasy,
I gaze upon thy glory and grow bright:
As if irradiate with beholden light;
As if the immortal that remains of thee
Attune me to thy spirit's harmony,
Breathing serene resolve and tranquil might,
Revealed appear thy silent thoughts of youth,
As if to consciousness, and all that view
Prophetic, of the heritage of truth
To thy majestic years of manhood due:
Darkness and error fleeing far away,
And the pure mind enthroned in perfect day.

As Fourier analysts, what can we add to this panegyric to Fourier? We were tempted to stop there, but decided to add some mathematics in order to amplify the poetry and thus emphasize the power of the rhyme. "There will be rhymes and mathematics..." as the famous Russian bard Vladimir Vysotsky sang in one of his songs. To this end, we chose a set of topics that satisfy two important criteria. First, we present mathematics that we actually understand. While this seemingly natural obstacle did not hold back every author in the history of written word, we decided that it is an important factor for us. Second, we emphasize those topics where the relationship between analytic and geometric reasoning is on full display, both in terms of techniques and motivations. The study of properties of decay of the Fourier transform of suitably regular functions and measures is a major organizing theme of this text.

Every time a paper or a book has more than one author one can legitimately ask whether the final product is better as a result. In other words, what is the purpose of this particular collaboration, besides providing ample excuses to discuss chess, literature and (alas) international politics among the authors? On a more serious note, the decay of the Fourier transform is a deep and extensive field with a variety of applications and connections to many areas of mathematics. Combining our efforts and interweaving our expertise allowed us to broaden the book and make it more appealing to a wider audience of scientists.

Though modern means of communication allow people to feel as if they are in the same room, much of our work – indeed the most important part of it – was done when we had opportunities to devote our time together to the actual writing of the book that followed our individual research and thought. First of all our home universities should be mentioned for giving us such opportunities. These were and are University of Missouri, Columbia, and University of Rochester, New York, both USA, and Bar-Ilan University, Ramat-Gan, Israel. These intensive periods of writing required extra energy in the form of food, drink and intellectual and emotional support. Our wonderful spouses played a crucial role in this process and our gratitude to them is immense.

During our careers we met many people who influenced our mathematics in variety of direct and subtle ways. Any attempt to list them all would inevitably lead to omissions. Having said that, here is a list of people who had a significant impact on our understanding of mathematics related to the subject matter of this book.

It is our pleasure to express our sincere gratitude to Marshall Ash, Eduard Belinsky, Luca Brandolini, Michael Christ, Leonardo Colzani, Laura De Carli, Burak Erdogan, Hans Feichtinger, Sandor Fridli, Michael Ganzburg, John Garnett, Daryl Geller, Allan Greenleaf, Steve Hofmann, Olga Kuznetsova, Michael Lacey, Ferenc Móricz, Pertti Mattila, Fedor Nazarov, Yuri Nosenko, Alexander Olevskii, Konstantin Oskolkov, Anatoly Podkorytov, Fulvio Ricci, Misha Rudnev, Stefan Samko, Eric Sawyer, Andreas Seeger, Maria Skopina, Christopher Sogge, Elias Stein, Sergey Tikhonov, Giancarlo Travaglini, Walter Trebels, Sergei Treil, Roald Trigub, Ulrich Stadtmüller, Sasha Volberg, Steven Wainger, Vladimir Yudin, and Georg Zimmermann.

We would like to express our deepest gratitude to the Birkhäuser Publishing House in general and Sylvia Lotrovsky and Thomas Hempfling in particular. The patience they have exhibited during this process boggles the mind. During several long stretches of time when the authors themselves doubted whether the book would ever come into being, the editors' faith proved absolutely unprecedented.

Alex Iosevich
Elijah Liflyand

Part 0

Preliminaries

Introduction

This is a rather informal introduction, designed to describe the ideas and points of view of this book, rather than precise definition and calculations. All issues described below are treated in a precise matter later in the book. Our purpose is to show how various concepts arise, why they are important, and how they fit together. In other words, one should view this introduction as an informal microcosm of the book which can be read separately or as a launch pad into the rest of the treatise.

We also wish to make a rather obvious disclaimer before we plunge into this introduction. The book is not meant to be an exhaustive treatise on the asymptotic behavior of the Fourier transform. Any such attempt would be both arrogant and foolish. Instead we present a segment of this rich and beautiful theory mostly dealing with the properties of the Fourier transform as it arises in the study of local geometric properties of sub-manifolds of the Euclidean space and the associated functions. We can only regret that many interesting related topics are not touched upon for one or another reason; we mention just as an example some relations between the decay of the Fourier transform and (pointwise) multipliers, see, *e.g.*, the monograph by Maz'ya and Shaposhnikova [143, Ch. 4, 4.4.2, Th. 4.4.3].

The Fourier transform is a ubiquitous object in modern analysis, physics and engineering. Far from being a "trick" or a "tool", it is rather a fundamental operation which relates the spacial properties of a function with its frequency behavior. The notion and applications of the Fourier transform are intimately tied to the Plancherel theorem, and this is where we begin. Let $f \in L^p(\mathbb{R}^d)$, $p \geq 1$, be the equivalence class of functions satisfying

$$\|f\|_{L^p(\mathbb{R}^d)} = \left(\int_{\mathbb{R}^d} |f(x)|^p dx \right)^{\frac{1}{p}} < \infty.$$

Define the Fourier transform by the formula

$$\widehat{f}(\xi) = \int_{\mathbb{R}^d} e^{-2\pi i x \cdot \xi} f(x)\, dx.$$

The Fourier transform of an arbitrary L^1 function may tend to 0 arbitrarily slowly at infinity. We are interested in quantifying this rate of decay and expressing it in terms of natural and easily computable properties of f, especially those arising in connection with geometric problems.

The most basic formula, which represents raison d'être of the Fourier transform, is the Plancherel's formula which says that

$$\|f\|_{L^2(\mathbb{R}^d)} = \|\widehat{f}\|_{L^2(\mathbb{R}^d)}.$$

This formula suggests very strongly that regularity properties of f directly impact the decay rate of $\widehat{f}$. This, in many ways, is what this book is about.

Let us continue by considering a smooth function ϕ, identically equal to 1 in the unit ball and vanishing outside the ball of radius 2. Let $\phi_{\epsilon,2}(x) = \epsilon^{-\frac{d}{2}}\phi(\frac{x}{\epsilon})$. Observe that

$$\|\phi_{\epsilon,2}\|_{L^2(\mathbb{R}^d)}$$

is independent of ϵ. What interests us is the fact that $\phi_{\epsilon,2}$ is concentrated in a ball of radius ϵ and is equal to approximately $\epsilon^{-\frac{d}{2}}$ in that ball. On the other hand,

$$|\widehat{\phi}_{\epsilon,2}(\xi)| = \epsilon^{\frac{d}{2}}|\widehat{\phi}(\epsilon\xi)| \leq \epsilon^{\frac{d}{2}}C_N(1+|\epsilon\xi|)^{-N}$$

for any $N > 0$. This function is "concentrated" in a ball of radius ϵ^{-1} and is equal to approximately $\epsilon^{\frac{d}{2}}$ in that ball. What we have just illustrated is a special case of a theorem that frequently goes under the heading of a Heisenberg Uncertainty Principle, which says, informally speaking, that the Fourier transform of a sharp short signal is long and shallow.

The example in the previous paragraph gives us an excellent means to continue our investigation of the relationship between properties of the function and the decay rate of its Fourier transform. Let us redefine $\phi_{\epsilon,2}$ slightly by setting $\phi_{\epsilon,1}(x) = \epsilon^{-d}\phi(\frac{x}{\epsilon})$. Suppose in addition that $\int_{\mathbb{R}^d}\phi(x)dx = 1$. This function converges to the point mass at the origin in the sense that if g is an integrable function,

$$\int \phi_{\epsilon,1}(x)g(x)dx \to g(0) \tag{0.1}$$

for almost every $x \in \mathbb{R}^d$. On the other hand,

$$|\widehat{\phi}_{\epsilon,1}(\xi)| = |\widehat{\phi}(\epsilon\xi)| \to 1 \tag{0.2}$$

as $\epsilon \to 0$.

Let us turn our attention to a "thicker" function. Let f be the characteristic function of the cube $Q_d = [-1,1]^d$. Then

$$\widehat{f}(\xi) = \prod_{j=1}^{d} \frac{\sin(\pi\xi_j)}{\pi\xi_j}.$$

When ξ is away from the coordinate axis, $|\widehat{f}(\xi)| \leq (\pi|\xi|)^{-d}$. However, when $\xi = \xi_j$, all we can say is that $\widehat{f}(\xi) \leq (\pi|\xi|)^{-1}$. This obnoxious behavior does not take place "often". We can formalize this idea by recalling Plancherel which tells us that

$$2^d = \int_{\mathbb{R}^d} |f(x)|^2 dx = \int_{\mathbb{R}^d} \left|\widehat{f}(\xi)\right|^2 d\xi,$$

which suggests that

$$\left(\frac{1}{R^d} \int_{R \leq |\xi| \leq 2R} \left|\widehat{f}(\xi)\right|^2 d\xi\right)^{\frac{1}{2}}$$

should tend to 0 as $R \to \infty$, the question being how fast. Using the divergence theorem, it is not difficult to check that if σ_{Q_d} is the Lebesgue measure on the boundary of Q_d, then

$$\left(\frac{1}{R^d} \int_{R \leq |\xi| \leq 2R} |\widehat{\sigma}_{Q_d}(\xi)|^2 d\xi\right)^{\frac{1}{2}} \lesssim R^{-\frac{d-1}{2}},$$

where here and for the remainder of the book, $a \lesssim b$ means that there exists a positive constant C, independent of the essential parameters, such that $a \leq Cb$.

We just saw in (0.2) that for a single point, the Fourier transform does not decay at all, even on average, whereas for a $d-1$-dimensional object, the boundary of the unit cube, the L^2 average of the Fourier transform on the scale R decays like $R^{-\frac{d-1}{2}}$. Let us try to understand the sense in which this is a general phenomenon. The fact that $\mathbb{R}^d$ is d-dimensional is captured locally by the fact that the function $|x|^{-\gamma}$ is integrable near the origin as long as $\gamma < d$. Let us use this idea to say, for the moment, that $E \subset \mathbb{R}^d$ is α-dimensional if there exists a Borel measure μ supported on E such that, for every $\beta < \alpha$,

$$\int\int |x-y|^{-\beta} d\mu(x)d\mu(y) < \infty. \tag{0.3}$$

Using the Fourier transform, we rewrite (0.3) in the form

$$C_\beta \int |\widehat{\mu}(\xi)|^2 |\xi|^{-d+\beta} d\xi < \infty.$$

It follows that

$$\left(\frac{1}{R^d} \int_{R \leq |\xi| \leq 2R} |\widehat{\mu}(\xi)|^2 d\xi\right)^{\frac{1}{2}} \lesssim R^{-\frac{\beta}{2}},$$

so the phenomenon we discovered by looking at the point and the boundary of the square is, indeed, "dimensional" in nature. Is dimension the only determining factor in the behavior of the Fourier transform? We can see that the answer is no in a variety of ways, but perhaps the simplest is the following. Let σ_{flat} denote the Lebesgue measure on the line $\{(s,s) : 0 \leq s \leq 1\}$ and let σ_{curved} denote the Lebesgue measure on the parabola $\{(s,s^2) : 0 \leq s \leq 1\}$. We have

$$\widehat{\sigma}_{\text{flat}}(\xi) = \int_0^1 e^{-2\pi i(s\xi_1 + s\xi_2)} ds,$$

and

$$\widehat{\sigma}_{\text{curved}}(\xi) = \int_0^1 e^{-2\pi i(s\xi_1+s^2\xi_2)}ds.$$

Taking $\xi_1 = \xi_2$ makes it clear that $\widehat{\sigma}_{\text{flat}}(\xi)$ does not in general tend to 0 at infinity, whereas a simple calculation with Gaussians shows that $|\widehat{\sigma}_{\text{curved}}(\xi)| \lesssim |\xi|^{-\frac{1}{2}}$. What is the difference between the two measures? In the language of elementary analytic geometry, the parabola $\{(s,s^2) : s \in [0,1]\}$ has everywhere non-vanishing curvature, whereas the curvature on the line segment $\{(s,s) : s \in [0,1]\}$ is zero everywhere.

So far we have encountered two types of averages for the Fourier transform:

$$\text{Solid average: } \left(\frac{1}{R^d}\int_{R\leq|\xi|\leq 2R} |\widehat{f}(\xi)|^2 d\xi\right)^{\frac{1}{2}} \lesssim |\xi|^{-\gamma_a}$$

for an appropriate $\gamma_a \geq 0$, and

$$\text{Pointwise estimate: } |\widehat{f}(\xi)| \lesssim |\xi|^{-\gamma_p}$$

for some appropriate $\gamma_p \geq 0$.

We have seen that, roughly speaking, the first estimate only depends on "dimensional" properties of f, measured by the integrability condition (0.3). On the other hand, we have seen that the second estimate depends very much on the local geometry. Another way of saying this is that the solid average reflects mainly the analytic aspects of Fourier decay, whereas the pointwise estimate is fundamentally dependent on the underlying geometry. In practice, the dichotomy is not quite so rigid; in fact the analysis and geometry interact to various extents in both types of estimates, yet the general picture is effectively described in this fashion.

We have absolutely no desire to artificially engineer intermediate scenaria, but there are one or two that very much deserve our attention. Let P be a polygon with infinitely many sides such that the normals to these sides form a sequence $\{n^{-a}, n = 1,2,\dots\}$, $1 < a < \infty$. Let σ_P denote the Lebesgue measure on P. By taking ξ to be in the direction normal to one of the sides in P, it is not difficult to see that $\widehat{\sigma}_P$ does not in general tend to 0 as $|\xi| \to \infty$. It is of course true that

$$\left(\frac{1}{R^d}\int_{R\leq|\xi|\leq 2R} |\widehat{f}(\xi)|^2 d\xi\right)^{\frac{1}{2}} \lesssim R^{-\beta} \tag{0.4}$$

with any $\beta < 1$, since, after all, P is one-dimensional. We can say more, however. We shall see that

$$\int_0^{2\pi} |\widehat{\sigma}_P(R\cos(\theta), R\sin(\theta))| d\theta \lesssim R^{-1+\frac{1}{2}\frac{1}{2+a}}, \tag{0.5}$$

and

$$\left(\int_0^{2\pi} |\widehat{\sigma}_P(R\cos(\theta), R\sin(\theta))|^2 d\theta\right)^{\frac{1}{2}} \lesssim R^{-\frac{1}{2}}. \tag{0.6}$$

One can check directly that (0.6) implies (0.4). On the other hand, the estimate (0.5) demonstrates a whole new feature of the Fourier transform. This estimate shows that the behavior of the Fourier transform is tied not only to local, but also micro-local properties of the set in question. More precisely, the estimates depend on the Minkowski dimension of the set of vectors normal to the boundary.

We shall also see that the estimate like (0.6) need not hold for general one-dimensional measures. The condition under which (0.6) holds, even for one-dimensional measures, very much remains an open question.

The notion of spherical decay is extremely useful in the study of the distribution of lattice points inside convex domains. More precisely it allows us to study questions of the following type. Let K_θ denote a convex body rotated by an angle $\theta \in SO(d)$. Let $N_\theta(t) = \#\{\mathbb{Z}^d \cap tK_\theta\} = t^d|K_\theta| + E_\theta(t)$. The problem, motivated by questions in number theory, spectral theory and other areas of mathematics, is to find the best rate of growth of $|E_\theta(t)|$ for almost every θ.

This connection with analytic number theory is neither singular nor accidental. Throughout this book we will explore connections between Fourier analytic inequalities and number theoretic concepts and results. This symbiosis often provides insight into the numerology behind Fourier analytic estimates. For example, if K is a polygon, its boundary has a finite number of normals, and it is relatively easy to rotate it in such a way that the boundary of the dilate tK_θ encounters very few lattice points in $\mathbb{Z}^d$. This results in a small error term E_θ described in the previous paragraph. Using the Poisson Summation Formula, properties of E_θ are connected, in a direct way, to properties of $|\widehat{\chi}_{K_\theta}|$ and its L^1 averages

$$\int_0^{2\pi} |\widehat{\chi}_{K_\theta}(tk)|\, d\theta.$$

Precisely this line of thinking led the authors of [44] to conjecture estimates related to (0.5). We emphasize that in this case a variant of the classical lattice point problem was not simply an area of application of the given class of Fourier analytic estimates. Indeed, elementary number theoretic reasoning provided the necessary inspiration and geometric intuition that brought these analytic estimates into being.

Another notion of average decay arises in the study of distance sets. The basic problem, due to Falconer (and earlier by Erdős in a discrete setting), is to determine whether Hausdorff dimension greater than $\frac{d}{2}$ for a compact set E guarantees that the set of distances

$$\Delta(E) = \left\{\sqrt{(x_1 - y_1)^2 + \cdots + (x_d - y_d)^2} : x, y \in E\right\}$$

has positive Lebesgue measure. One approach to the problem is to try to obtain a good upper bound for the multiplicity function

$$m_\epsilon(t) = \mu \times \mu\{(x,y) \in E \times E : t - \epsilon \leq |x-y| \leq t+\epsilon\},$$

where μ is a Borel measure on E. This, however, is very difficult. Attempts to control m_ϵ in the L^2 norm, in the appropriate sense, lead to the following integral discovered by P. Mattila:

$$\int_1^\infty \left(\int_{S^{d-1}} |\widehat{\mu}(R\omega)|^2 d\omega \right)^2 R^{d-1} dR, \tag{0.7}$$

which can be viewed as an average of the Fourier transform intermediate between the "solid" average (0.4) and a spherical average of the form

$$\left(\int_{S^{d-1}} |\widehat{\mu}(R\omega)|^2 d\omega \right)^{\frac{1}{2}} \lesssim R^{-\beta}, \tag{0.8}$$

with an appropriate value of β, since (0.6) and (0.8) can be combined to yield an estimate on (0.7). If one wishes to study the distance set problem corresponding to the distance induced by a symmetric convex body K with a smooth boundary, then (0.8) takes the form

$$\left(\int_{\partial K^*} |\widehat{\mu}(R\omega)|^2 d\omega \right)^{\frac{1}{2}} \lesssim R^{-\beta}, \tag{0.9}$$

where $K^* = \{x \in \mathbb{R}^d : x \cdot y \leq 1 \ \forall \ y \in K\}$, the dual body of K.

Again number theory is lurking in the background. Suppose that (0.9) holds with the optimal exponent $\beta = s$, the Hausdorff dimension of the set E where μ is supported. Choose a special case where E is the scaled lattice $\frac{1}{q}(\mathbb{Z}^d \cap [0,q]^d)$ thickened by q^{-2}. Let $\mu_q(x)$ be the characteristic function of this scaled thickened lattice, normalized to achieve total unit mass. While the support of this measure is a set of positive Lebesgue measure, it is asymptotically, as $q \to \infty$, a measure on a set of Hausdorff dimension $\frac{d}{2}$. We shall see in the sequel that the inequality (0.9) with $\beta = \frac{d}{2}$, $d \geq 4$, would imply that for any symmetric convex body K with a smooth boundary,

$$\#\left\{ k \in \mathbb{Z}^d : R - \frac{1}{R} \leq \|k\|_K \leq R + \frac{1}{R} \right\} \leq CR^{d-2}.$$

In other words, inequality (0.9) would imply that a thin neighborhood of the dilated boundary of a smooth convex body contains no more lattice points than does the sphere of the same dimension. Even in the case of the sphere, this is a deep classical result due to Landau. In the case of general convex bodies with smooth boundaries this is an important wide open problem. This not only provides a connection between the distance set problems and their Fourier analytic apparatus and number theory, but also clearly suggests the level of depth that would be required to solve the analytic estimates under consideration.

Up to this point our discussion has centered around size estimates of the Fourier transform, whether pointwise or average. However, precise asymptotic behavior is also extremely important and useful. A beautiful formula attributed to Herz [94] on this side of the Iron Curtain, and to Gelfand, Graev, and Vilenkin on the other side, tells us that if χ_K is the characteristic function of a convex body $K = \{x : \rho(x) \leq 1\}$, symmetric about the origin and having everywhere non-vanishing Gaussian curvature on the boundary, then

$$\widehat{\chi}_K(\xi) = C|\xi|^{-\frac{d+1}{2}} \sin\left(2\pi\left(\rho^*(\xi) + \frac{d-1}{8}\right)\right) + O(|\xi|^{-\frac{d+3}{2}}). \tag{0.10}$$

Using estimates of this type we shall study the structure of zeroes of the Fourier transform of characteristic functions of convex sets and tie the theory of the Fourier transform to geometric problems in partial differential equations such as the Pompeiu conjecture and Erdős type problems in combinatorial geometry. For example, we shall see that the formula can be used to show that if $K \subset \mathbb{R}^d$ is as above, and $\{e^{2\pi i x\cdot a}\}_{a\in A}$ is an orthogonal family in the sense that

$$\int_K e^{2\pi i x\cdot(a-a')}dx = 0 \ for \ a \neq a',$$

then the cardinality of A is finite if $d \equiv 1(4)$. If $d \equiv 1(4)$, the set A may be infinite, but in that case it must be a subset of a line.

The formula (0.10) above is a special case of an asymptotic development of the Fourier transform of a general function. Why should one be interested in such an object? Let us describe an important example. Consider truncated Fourier series of the form

$$S_N^\lambda f(x) = \sum_{k\in\mathbb{Z}^d} \lambda\left(\frac{k}{N}\right)\widehat{f}(k)e^{2\pi i x\cdot k},$$

where λ is a suitable function, not necessarily with compact support. The study of convergence properties of $S_N^\lambda f(x)$ leads one to consider (generalized) Lebesgue constants

$$L_N = \int_{\mathbb{T}^d}\left|\sum_{k\in\mathbb{Z}^d}\lambda\left(\frac{k}{N}\right)e^{2\pi i x\cdot k}\right|dx.$$

For many choices of the function λ one can show that

$$L_N = \int_{|x|\leq N}|\widehat{\phi}(x)|dx + \text{ suitably small error},$$

and this is where asymptotics of Fourier transforms of general functions comes into play. In many cases it is important that the transformed function be of bounded variation, usually in dimension one but in certain cases generalized in several dimensions.

Many important classical examples of functions λ above are radial. Perhaps the most celebrated case is the one where

$$\lambda(x) = \left(1 - |x|^2\right)_+^\alpha,$$

known as the Bochner–Riesz kernel. See, for example, [180] and the references therein.

Such examples illustrate the need for asymptotic expansions of Fourier transforms of radial function, a subject that we treat in considerable detail in the sequel.

The matter of the last few paragraphs is what we treat as analytic aspects of the decay of the Fourier transform, while all the preceding matter that leads to these we mostly treat as geometric aspects. However our intuition, rather than precise argument, has resulted in the reverse order of presentation; moreover, the two types of aspects almost always have something in common.

We sincerely hope that this introduction has left you with a vaguely unsatisfactory feeling of having just scratched the surface, and a desire to attain a tighter grasp of the subject matter on the level of numerous and often painful details. This brings us into the main body of this book where the adventure begins in earnest.

Chapter 1

Basic Properties of the Fourier Transform

In this chapter we define the Fourier transform and describe its basic properties. Since this part of the book is quite standard, we go through the material quickly with an eye on developments in the subsequent chapters.

1.1 L^1-theory

Let $f \in L^1(\mathbb{R}^d)$. Define the Fourier transform of f by the formula

$$\widehat{f}(\xi) = \int_{\mathbb{R}^d} e^{-2\pi i x\cdot\xi} f(x)dx, \tag{1.1}$$

where

$$x \cdot \xi = x_1\xi_1 + \cdots + x_d\xi_d.$$

Theorem 1.2. *The mapping $f \to \widehat{f}$ is a bounded linear map from $L^1(\mathbb{R}^d)$ to $L^\infty(\mathbb{R}^d)$, $\widehat{f}$ is continuous, $\lim_{|\xi|\to\infty} \widehat{f}(\xi) = 0$, and*

$$\|\widehat{f}\|_{L^\infty(\mathbb{R}^d)} \le \|f\|_{L^1(\mathbb{R}^d)}. \tag{1.3}$$

The first two properties follow directly from the definition of the Fourier transform. In order to prove the third claim, known as the Riemann–Lebesgue lemma, first prove it for the characteristic (indicator) function of a cube by an explicit calculation, and then apply a limiting argument. The estimate (1.3) follows by the definition of the Fourier transform.

It is reasonable to ask at this point how fast $\widehat{f}(\xi)$ goes to 0 as $|\xi| \to \infty$. It is known that an arbitrary rate of convergence to 0 is possible. Indeed, a classical result due to Pólya (see, *e.g.*, [139] or [203]) says that each even, convex and monotone decreasing to zero function on $[0, \infty)$ is the Fourier transform of an integrable function. More sophisticated examples of that kind can be found in [41].

Theorem 1.4. *Let* $t_a f(x) = f(x-a)$. *Then*

$$\widehat{t_a f}(\xi) = e^{2\pi i a\cdot\xi}\widehat{f}(\xi).$$

Similarly, the Fourier transform of $e^{2\pi i x\cdot a}f(x)$ *is* $\widehat{f}(\xi - a)$.

With this simple example behind us, it is reasonable to ask how the Fourier transform behaves under the influence of general linear transformations.

Theorem 1.5. *Let* T *be a non-singular complex-valued linear map from* $\mathbb{R}^d$ *to itself and define* $f_T(x) = f(T(x))$. *Then*

$$\widehat{f_T}(\xi) = |T|^{-1}\widehat{f}((T^{-1})^*(\xi)),$$

where T^* *denotes the adjoint of* T *and* $|T|$ *is the determinant.*

To prove this result, observe that

$$\begin{aligned}\widehat{f_T}(\xi) &= \int_{\mathbb{R}^d} e^{-2\pi i x\cdot\xi} f(Tx)\,dx = |T|^{-1}\int_{\mathbb{R}^d} e^{-2\pi i T^{-1}x\cdot\xi} f(x)\,dx \\ &= |T|^{-1}\int_{\mathbb{R}^d} e^{-2\pi i x\cdot(T^{-1})^*\xi} f(x)\,dx = \widehat{f}((T^{-1})^*)\end{aligned}$$

since $Tx\cdot y = x\cdot T^*y$ by definition of the transpose.

In particular, Theorem 1.5 gives that if $f(x)$ is radial (depending only on $|x|$), then $\widehat{f}$ is radial too. More precisely, it is represented as a Hankel transform as follows.

Theorem 1.6. *Let* $t^{d-1}f_0(t)$ *be integrable on* $(0,+\infty)$. *Then for its radial extension* $f(x) = f_0(|x|)$,

$$\widehat{f}(\xi) = 2\pi\int_0^{+\infty} f_0(t)(|\xi|t)^{1-\frac{d}{2}} J_{\frac{d}{2}-1}(2\pi|\xi|t)t^{d-1}dt. \tag{1.7}$$

This formula is referred to in [30] as the Cauchy–Poisson formula.

We now turn towards the issue of the behavior of the Fourier transform under the influence of differentiation.

Theorem 1.8. *Let* f_j *denote the partial derivative of* f *with respect to* x_j. *Suppose that* $f_j \in L^1(\mathbb{R}^d)$. *Then*

$$\widehat{f_j}(\xi) = 2\pi i\xi_j\widehat{f}(\xi).$$

Similarly, if $x_j f \in L^1(\mathbb{R}^d)$,

$$-2\pi i\widehat{x_j f}(\xi) = \frac{\partial\widehat{f}}{\partial\xi_j}(\xi).$$

Both identities follow easily from the definition using integration by parts.

1.2 Convolution

Another basic operation with respect to the Fourier transform is the convolution

$$f * g(x) = \int_{\mathbb{R}^d} f(x-y)g(y)\,dy,$$

which is well defined for $L^1(\mathbb{R}^d)$ functions by Fubini's theorem. Its relationship to the Fourier transform is given by the following basic calculation.

Theorem 1.9. *Let $f, g \in L^1(\mathbb{R}^d)$. Then*

$$\widehat{f * g}(\xi) = \widehat{f}(\xi)\widehat{g}(\xi).$$

To see this, write

$$\begin{aligned}\widehat{f * g}(\xi) &= \int_{\mathbb{R}^d} e^{-2\pi i x\cdot\xi} \int_{\mathbb{R}^d} f(x-y)g(y)\,dy\,dx \\ &= \int_{\mathbb{R}^d}\int_{\mathbb{R}^d} e^{-2\pi i(x-y)\cdot\xi} e^{-2\pi i y\cdot\xi} f(x-y)g(y)\,dx\,dy \\ &= \int_{\mathbb{R}^d} e^{-2\pi i u\cdot\xi} f(u)\,du \cdot \int_{\mathbb{R}^d} e^{-2\pi i v\cdot\xi} f(v)\,dv = \widehat{f}(\xi)\widehat{g}(\xi).\end{aligned}$$

An important tool is Young's inequality for convolution (see, *e.g.*, [181, Ch. V, §1]):

Theorem 1.10. *If $\varphi \in L^r(\mathbb{R})$ and $\psi \in L^q(\mathbb{R})$, then for $\frac{1}{r} + \frac{1}{q} = \frac{1}{p} + 1$, $1 \le p, q, r \le \infty$,*

$$\|\varphi * \psi\|_p \le \|\varphi\|_r \|\psi\|_q. \tag{1.11}$$

1.3 L^2-theory

While this may appear slightly paradoxical, we begin the section on L^2-theory by defining a set of functions which are rather more regular.

Definition 1.12. We say that $\phi \in C^\infty(\mathbb{R}^d)$ belongs to $\mathcal{S}(\mathbb{R}^d)$ if

$$\sup_{x\in\mathbb{R}^d} |x^\gamma \partial^\alpha \phi(x)| < \infty$$

for all multi-indices γ and α. Here, and throughout the book, $x^\gamma = x_1^{\gamma_1} \cdots x_d^{\gamma_d}$, and $\partial^\alpha \phi$ denotes the partial derivative of order $\alpha_1 + \cdots + \alpha_d$, with respect to $x_1, \ldots, x_d$ respectively.

It follows easily from Theorem 1.8 that the Fourier transform maps $\mathcal{S}(\mathbb{R}^d)$ to itself. Moreover, we have the following fundamental fact.

Theorem 1.13. *The Fourier transform is an isomorphism of $\mathcal{S}(\mathbb{R}^d)$ into itself whose inverse is given by the formula*

$$f(x) = \int_{\mathbb{R}^d} e^{2\pi i x\cdot\xi} \widehat{f}(\xi)\, d\xi.$$

To see this, we need the following basic calculations.

Lemma 1.14. *Let $f, g \in L^1(\mathbb{R}^d)$. Then*

$$\int_{\mathbb{R}^d} \widehat{f}(x) g(x)\, dx = \int_{\mathbb{R}^d} f(x)\widehat{g}(x)\, dx.$$

The proof of this is immediate by Fubini. We also need to know that, roughly speaking, the Fourier transform of a Gaussian is a Gaussian.

Lemma 1.15. *Let $\gamma(x) = e^{-i\pi x^2}$. Then*

$$\widehat{\gamma}(\xi) = \gamma(\xi).$$

The proof is by completing the square and changing the contour of integration.

Using Lemma 1.15, it is not difficult to derive the following basic relation, known as the Fourier inversion formula.

Theorem 1.16. *Suppose that $f \in L^1$, and assume that $\widehat{f}$ is also in L^1. Then for almost every x,*

$$f(x) = \int_{\mathbb{R}^d} e^{2\pi i x\cdot\xi} \widehat{f}(\xi)\, d\xi.$$

Perhaps the most fundamental result in Fourier analysis is Plancherel's theorem.

Theorem 1.17. *If $u, v \in \mathcal{S}$, then*

$$\int_{\mathbb{R}^d} \widehat{u}(x)\overline{\widehat{v}}(x)\, dx = \int_{\mathbb{R}^d} u(x)\overline{v}(x)\, dx.$$

Moreover, there is a unique operator $\mathcal{F} : L^2(\mathbb{R}^d) \to L^2(\mathbb{R}^d)$ such that $\mathcal{F}f = \widehat{f}$ when $f \in \mathcal{S}(\mathbb{R}^d)$. This operator is unitary and $\mathcal{F}f = \widehat{f}$ when $f \in L^1 \cap L^2$.

1.4 Summability

Observe that Theorem 1.16 holds true under the very restrictive assumption that $\widehat{f}$ is in L^1. If it is not the case, summability is a typical tool to restore a function from its Fourier transform. Among various types of summability, the one by multipliers is the most common. The corresponding linear means are defined as

$$\int_{\mathbb{R}^d} \lambda\left(\frac{\xi}{R}\right) \widehat{f}(\xi) e^{2\pi i x\cdot\xi}\, dx,$$

where λ is a (multiplier) function. Some of such functions became widely known and used, see, *e.g.*, [63, Ch. 1, §9]. What is of crucial importance for us is that summability and approximation properties of the above linear means essentially depend upon the behavior of the Fourier transform of λ; the best source for the study of this aspect is [203]. One can see from this book that a similar situation takes place in the case of Fourier series. A counterpart of the above linear means for the Fourier series of a 1-periodic in each variable function f, defined on the torus $\mathbb{T}^d = (-\frac12, \frac12]^d$, looks like

$$\sum_{k\in\mathbb{Z}^d} \lambda_{k,N} \widehat{f}(k) e^{2\pi i x\cdot k},$$

where

$$\widehat{f}(k) = \int_{\mathbb{T}^d} f(x) e^{-2\pi i x\cdot k}\, dx$$

is the kth Fourier coefficient of f and $\lambda_{k,N}$ is a sequence of multipliers. When f is continuous or just integrable, summability properties are related to the behavior of the sequence of Lebesgue constants

$$\int_{\mathbb{T}^d} \left| \sum_{k\in\mathbb{Z}^d} \lambda_{k,N} e^{2\pi i x\cdot k} \right| dx;$$

a detailed overview of this subject is given in [125]. It is shown here (see also [203] and references therein) that the most interesting case is when

$$\lambda_{k,N} = \lambda\left(\frac{k}{N}\right),$$

where it is natural to take λ at least continuous to be defined at each point $\frac{k}{N}$. Such means are

$$L_N^\lambda(f;x) = \sum_{k\in\mathbb{Z}^d} \lambda\left(\frac{k}{N}\right) \widehat{f}(k) e^{2\pi i k\cdot x}. \tag{1.18}$$

The most known among such functions λ is $(1-|x|^2)^\alpha$ which defines the so-called Bochner–Riesz means of order α, introduced in [30] (for numerous important problems where the Bochner–Riesz means are involved, see, *e.g.*, [181] and [61]). Again, much clarifies itself from the behavior of the Fourier transform of λ. In fact, many of the results of the first part of this book are inspired or obtained in order to establish certain summability and/or approximation properties of the linear means of Fourier series.

1.5 Poisson summation formula

The Poisson summation formula was discovered by Siméon Denis Poisson and is sometimes called Poisson resummation. A typical form of the Poisson summation formula for integrable functions is given in [181, Ch. VII, Th. 2.4].

Theorem 1.19. *Suppose $f \in L^1(\mathbb{R}^d)$. Then the series*

$$\sum_{m\in\mathbb{Z}^d} f(x+m) \tag{1.20}$$

converges in the norm $L^1(\mathbb{T}^d)$. The resulting function in $L^1(\mathbb{T}^d)$ has the Fourier expansion

$$\sum_{m\in\mathbb{Z}^d} \widehat{f}(m)e^{2\pi i x\cdot m}.$$

This means that $\{\widehat{f}(m)\}$ is the sequence of the Fourier coefficients of the L^1 function defined by the series (1.20), where, for any $\xi \in \mathbb{R}^d$, we have (1.1).

Under certain restrictions (see, *e.g.*, [181, Ch. VII, Cor. 2.6]) one has

$$\sum_{m\in\mathbb{Z}^d} f(x+m) = \sum_{m\in\mathbb{Z}^d} \widehat{f}(m)e^{2\pi i x\cdot m},$$

and, in particular,

$$\sum_{m\in\mathbb{Z}^d} f(m) = \sum_{m\in\mathbb{Z}^d} \widehat{f}(m).$$

Moreover, results are known which show that the Poisson summation characterize, in that or another sense, the Fourier transform (see [58] and [67]).

There are specific versions of the Poisson summation formula for functions with bounded variation; see, *e.g.*, [215, Ch. II, §13], [202, Lemma 2], or [135].

A good source for various versions of the Poisson summation formula and their applications is [140, Ch. X, §6].

Part 1

Analytic (and Geometric) Aspects

Chapter 2

Oscillatory Integrals

The method of stationary phase is the term typically applied to study of the integrals of the form

$$\int_{\mathbb{R}^d} e^{iRG(x)}\psi(x)\,dx$$

by studying properties of derivatives of the real or complex-valued phase function $G(x)$ on the support of the cut-off $\psi(x)$. There are many reasons, both pure and applied, to study such integrals, and there is no reasonable way to describe even a representative sample of the ideas and motivations involved. For us, the motivation mainly comes from the study of Fourier transforms of measures supported on surfaces in Euclidean space possessing various degrees of smoothness. The applications of properties of Fourier transforms of such measures are found in partial differential equations, harmonic analysis, analytic number theory, integral geometry, geometric measure theory, and, in recent years, geometric combinatorics. We will make an effort in the coming pages to get across to the interested reader the variety and intricacy of applications of oscillatory integrals in different areas of mathematics, though, by necessity, we will only be scratching the surface.

2.1 The method of stationary phase

We begin with the analysis of the oscillatory integral of the form

$$I_f(R) = \int_a^b e^{iRf(x)}dx,$$

as R tends to infinity and f is a real-valued function. If $f(x)$ is constant, $I(R)$ does not decay at all. The point is that in order for $I_f(R)$ to behave well, the phase function $f(x)$ must change rapidly. To see this, note that

$$e^{iRf(x)} = \cos(Rf(x)) + i\sin(Rf(x)).$$

In order for $I_f(R)$ to be small, positive values of the cosine must cancel the negative values. For this to happen, at least one of the derivatives of $f(x)$ cannot be too small. For example, if $f(x) = x$,

$$|I_f(R)| = \left| \int_a^b e^{iRx}\,dx \right| = \left| \frac{1}{iR}\left(e^{iRb} - e^{iRa}\right) \right| \leq \frac{2}{R}.$$

With this calculation in mind, we state the classical van der Corput–Landau lemma.

Lemma 2.1. *Suppose that f is once differentiable on (a,b), $f'(x) \geq 1$, and f' is monotonic. Then*

$$|I_f(R)| \leq \frac{4}{R}.$$

Proof. To prove the lemma, write

$$\begin{aligned} I_f(R) &= \int_a^b e^{iRf(x)}\,dx \\ &= \int_a^b \frac{1}{iRf'(x)} \frac{d}{dx}\left(e^{iRf(x)}\right)\,dx \\ &= \left(\frac{e^{iRf(b)}}{iRf'(b)} - \frac{e^{iRf(a)}}{iRf'(a)} \right) - \int_a^b e^{iRf(x)} \frac{d}{dx}\left(\frac{1}{iRf'(x)}\right)\,dx \\ &= A + B. \end{aligned}$$

Since $f'(x) \geq 1$ on (a,b),

$$|A| \leq \frac{2}{R}.$$

Using the triangle inequality and the fact that f' is monotonic,

$$\begin{aligned} |B| &= \left| \int_a^b e^{iRf(x)} \frac{d}{dx}\left(\frac{1}{iRf'(x)}\right)\,dx \right| \\ &\leq \int_a^b \left| \frac{d}{dx}\left(\frac{1}{iRf'(x)}\right) \right|\,dx \\ &= \left| \int_a^b \frac{d}{dx}\left(\frac{1}{iRf'(x)}\right)\,dx \right| \\ &= \left| \left(\frac{1}{iRf'(b)} - \frac{1}{iRf'(a)} \right) \right| \leq \frac{2}{R}. \end{aligned}$$

It follows that

$$|I_f(R)| \leq |A| + |B| \leq \frac{4}{R},$$

as claimed. □

The natural question to ask at this point is what happens if oscillation is measured by higher-order derivatives. This suggests the possibility of an inductive argument, which is precisely what happens.

Lemma 2.2. *Let f be m times continuously differentiable on (a, b). Suppose that $f^{(m)}(x) \geq 1$. Then*

$$|I_f(R)| \leq C_m R^{-\frac{1}{m}},$$

where C_m depends only on m and does not depend on f, a or b.

Proof. To prove this lemma, assume that the result holds for $m-1$. If $f^{(m-1)}(x) \neq 0$ on (a, b), we are done, so we may assume that $f^{(m-1)}(x_0) = 0$ for some $x_0 \in (a, b)$. Since $f^{(m)}(x) \geq 1$, $f^{(m-1)}(x)$ is monotonic, as we noted above, so there exists at most one such x_0. Consider

$$\begin{aligned} I_f(R) &= \int_a^{x_0-\delta} e^{iRf(x)}dx + \int_{x_0-\delta}^{x_0+\delta} e^{iRf(x)}dx + \int_{x_0+\delta}^{b} e^{iRf(x)}dx \\ &= I_1 + I_2 + I_3. \end{aligned}$$

Using the fact that $e^{iRf(x)}$ has modulus one,

$$|I_2| \leq 2\delta.$$

To handle I_3 observe that

$$f^{(m-1)}(x) - f^{(m-1)}(x_0) = (x - x_0)f^{(m)}(c)$$

for some $c \in (x_0, x)$. It follows that

$$f^{(m-1)}(x) \geq \delta$$

on $(x_0 + \delta, b)$. By induction,

$$|I_3| = \left| \int_{x_0+\delta}^{b} e^{iRf(x)}dx \right| = \left| \int_{x_0+\delta}^{b} e^{iR\delta\frac{f(x)}{\delta}}dx \right| \leq C_{m-1}(R\delta)^{-\frac{1}{m-1}}.$$

Replacing $f(x)$ by $-f(x)$ and running the exact same argument shows that

$$|I_1| \leq C_{m-1}(R\delta)^{-\frac{1}{m-1}}.$$

We conclude that

$$|I_f(R)| \leq 2C_{m-1}(R\delta)^{-\frac{1}{m-1}} + 2\delta.$$

Choosing δ such that the two terms agree, we see that

$$|I_f(R)| \leq 4C_{m-1}^{\frac{m-1}{m}} R^{-\frac{1}{m}}.$$

This completes the proof with C_m given by the recursive relation $C_1 = 4$, which comes from Lemma 2.1, and $C_m = 4C_{m-1}^{\frac{m-1}{m}}$. □

A nearly immediate consequence of the van der Corput–Landau lemma is the following estimate on the Fourier transform of the Lebesgue measure on a smooth plane curve with everywhere non-vanishing curvature.

Lemma 2.3. *Let Γ be a smooth, compact plane curve with non-vanishing curvature. Let σ_Γ denote the Lebesgue measure on Γ. Then*

$$|\widehat{\sigma}_\Gamma(\xi)| \leq C|\xi|^{-\frac{1}{2}}.$$

Proof. By dividing the curve into finitely many pieces, we may assume that on each piece, Γ is a graph of a function $\gamma(t)$. We then use a smooth partition to express the measure on each piece of the curve as $\psi(t)dt$, where ψ is a smooth compactly supported function. After performing a translation and a rotation, we may assume that $\gamma(0) = \gamma'(0) = 0$. In these coordinates, the curvature function $\kappa(t)$ is given by the equation

$$\kappa(t) = \frac{\gamma''(t)}{(1 + (\gamma'(t))^2)^{\frac{3}{2}}}.$$

It follows that $\gamma''(t) \geq c > 0$. We must show that

$$|F_\gamma(\xi)| = \left| \int e^{-2\pi i(t\xi_1 + \gamma(t)\xi_2)} \psi(t)dt \right| \leq C|\xi|^{-\frac{1}{2}}.$$

First observe that if $|\xi_1| \geq C|\xi_2|$ for a sufficiently large constant C, then

$$|\widehat{F}_\gamma(\xi)| \leq c|\xi|^{-1}.$$

Indeed, let

$$\Phi(t) = 2\pi(t\xi_1 + \gamma(t)\xi_2).$$

Then

$$|\Phi'(t)| = 2\pi|\xi_1 + \gamma'(t)\xi_2| \geq |\xi_1|$$

if C is sufficiently large, since γ' is bounded above on a compact interval.

To take advantage of this, we integrate by parts just as we did in the proof of Lemma 2.1. We get

$$\begin{aligned}
\left| \int e^{-2\pi i(t\xi_1 + \gamma(t)\xi_2)} \psi(t)\, dt \right| &= \left| \int \frac{\psi(t)}{\Phi'(t)} \frac{d}{dt} \left(e^{-i\Phi(t)} \right) dt \right| \\
&= \left| \int e^{-i\Phi(t)} \frac{d}{dt} \left(\frac{\psi(t)}{\Phi'(t)} \right) dt \right| \leq \int \left| \frac{d}{dt} \left(\frac{\psi(t)}{\Phi'(t)} \right) \right| dt \\
&= \int \left| \frac{\Phi'(t)\psi'(t) - \psi(t)\Phi''(t)}{(\Phi'(t))^2} \right| dt \leq C|\xi_1|^{-1},
\end{aligned}$$

since $|\Phi'(t)| \geq |\xi_1|$ and $\Phi''(t)$ is bounded above on a compact interval.

If $|\xi_1| \geq C|\xi_2|$, then $|\xi_1|^{-1} \leq C'|\xi|^{-1}$. Therefore it remains to consider the case when $|\xi_1| \leq C|\xi_2|$. Fortunately, this puts us in the realm of Lemma 2.2. Indeed,

$$\int e^{-2\pi i(t\xi_1+\gamma(t)\xi_2)}\psi(t)\,dt = \int e^{-2\pi i\xi_2(t\xi_1\xi_2^{-1}+\gamma(t))}\psi(t)\,dt.$$

Since the second derivative of $t\xi_1\xi_2^{-1} + \gamma(t)$ is bounded from below by $c > 0$, by the curvature assumption, Lemma 2.2 implies that

$$\left|\int e^{-2\pi i\xi_2(t\xi_1\xi_2^{-1}+\gamma(t))}\psi(t)\,dt\right| \leq C|\xi_2|^{-\frac{1}{2}}.$$

Since we are in the regime where $|\xi_1| \leq C|\xi_2|$, the right-hand side above is bounded by $C|\xi|^{-\frac{1}{2}}$ and the proof is complete. □

For a more advanced application of the Stationary Phase Method, see, *e.g.*, calculations with Hardy's L-functions (see [91]) by Miyachi in [145].

Let us give a couple of versions of the Stationary Phase Method theorems in several dimensions. The first one can be found in [70].

Theorem 2.4. *For the integer $k \geq 1$ the following asymptotic formula is valid:*

$$\begin{aligned} Q_d(R) &= \int_{\mathbb{R}_+^{d-1}} \varphi(v)e^{iRS(v)}\,dv \\ &= (2\pi)^{\frac{d-1}{2}} R^{\frac{1-d}{2}} e^{i(RS(v_0)+\theta(v_0))} \\ &\quad \times |\det S''(v_0)|^{-\frac{1}{2}} \left(\varphi(v_0) + O(R^{-1})\right) \\ &\quad + R^{\frac{1-d}{2}} e^{iRS(v_0)} \sum_{j=1}^{k-1} a_j R^{-j} + O\left(R^{\frac{1-d}{2}-k}\right), \end{aligned} \tag{2.5}$$

where $v_0 = (v_1^0, v_2^0, \ldots, v_{d-1}^0)$ is a stationary point of S; S'' is the Hessian matrix of the second derivatives of S such that $S''(v_0) \neq 0$; $\theta(v_0)$ is a real number depending on $\det S''(v_0)$*; and a_j are some (complex) numbers.*

Let us give a somewhat different version, see [177, Ch. 1]. We consider the d-dimensional oscillatory integrals

$$I(R) = \int_{\mathbb{R}^d} a(R,x)e^{iRS(x)}\,dx, \qquad R > 0,$$

which involve C^∞ functions S and a, with S being real-valued and a having compact support. We will be working with phase functions S having non-degenerate critical points. Recall that x_0 is said to be a non-degenerate critical point if $\nabla S(x_0) = 0$ but $S''(x_0) \neq 0$.

Notice that non-degenerate critical points must be isolated since, by Taylor's theorem, near a non-degenerate critical point x_0,

$$\Delta S(x) = \frac{1}{2}\langle S''(x-x_0),(x-x_0)\rangle + O(|x-x_0|^3),$$

and hence

$$\nabla S(x) = S''(x-x_0) + O(|x-x_0|^2).$$

Finally, we shall say that S is a non-degenerate phase function if all of its critical points are non-degenerate. We shall work with amplitudes $a(R,x)$ whose x-support is contained in a fixed compact set, and we shall also require that

$$\left|\left(\frac{\partial}{\partial R}\right)^m \left(\frac{\partial}{\partial x}\right)^\gamma a(R,x)\right| \le C_{m\gamma}(1+R)^{-\alpha},$$

for all α and γ.

Theorem 2.6. *Suppose that a is as above, $S(0) = 0$, and 0 is a non-degenerate critical point of S. Then if $\nabla S(x) \neq 0$ on $\operatorname{supp} a(R,\cdot) \setminus \{0\}$,*

$$\left|\left(\frac{\partial}{\partial R}\right)^m I(R)\right| \le C_m(1+R)^{-\frac{d}{2}-m}.$$

Corollary 2.7. *We have*

$$\int_{\mathbb{R}^d} \eta(x)e^{iRS(x)}\,dx = R^{-\frac{d}{2}}(2\pi)^{\frac{d}{2}}e^{iRS(0)}\eta(0)|\det S''(0)|^{-\frac{1}{2}}e^{\frac{\pi i}{4}\operatorname{sign} S''} + O(R^{-\frac{d}{2}-1}),$$

if S has a non-degenerate critical point at 0 and η has small support.

Notice that, in this case, $\operatorname{sign} S''$ is constant on $\operatorname{supp}\eta$, so the first term on the right-hand side is well defined.

2.2 Erdélyi type results

The estimates we have carried out for $F_\gamma(\xi)$ in the previous section could not go through if the function ψ or certain of its derivatives had singularities at some point. To this end, we would like to have a possibility to deal with functions of type

$$e^{2\pi iyx}(x-a)^{\lambda-1}(b-x)^{\mu-1}.$$

Scrutiny of the Fourier integrals of such functions is due to A. Erdélyi (see, *e.g.*, [66], [57, §11]).

For any reasonable function f its repeated rth integral, $r = 1,2,\ldots$, over the interval $[a,x]$ is expressed as

$$\frac{1}{(r-1)!}\int_a^x (x-t)^{r-1}g(t)\,dt.$$

Here a is arbitrary, it will be convenient for us to integrate in the complex plane with $a = \infty i$. Writing

$$g(x) = e^{iyx}(x-a)^{\lambda-1},$$

with $y > 0$ and $0 < \lambda < 1$, we set

$$\begin{aligned}(K^r g)(x) &= \frac{1}{(r-1)!}\int_{\infty i}^{x}(x-\tau)^{r-1}g(\tau)\,d\tau \\ &= \frac{1}{(r-1)!}\int_{\infty i}^{x}(x-\tau)^{r-1}(\tau-a)^{\lambda-1}e^{iy\tau}\,d\tau,\end{aligned}$$

with integration along the ray in the complex plane going out from x to infinity parallel to the imaginary axis, $x \geq a$, and the principal value of $\tau - a)^{\lambda-1}$.

Substituting $\tau = x + it$, we obtain

$$(K^r g)(x) = \frac{e^{iyx - \frac{ri\pi}{2}}}{(r-1)!}\int_0^\infty e^{-yt}t^{r-1}(x-a+it)^{\lambda-1}\,dt.$$

In particular,

$$\begin{aligned}(K^r g)(a) &= \frac{e^{iya + \frac{\lambda\pi i}{2} - \frac{(r+1)\pi}{2}}}{(r-1)!}\int_0^\infty e^{-yt}t^{r+\lambda-2}\,dt \\ &= e^{iya + \frac{\lambda\pi i}{2} - \frac{(r+1)\pi}{2}}\,\frac{\Gamma(r+\lambda-1)}{(r-1)!y^{r+\lambda-1}}.\end{aligned} \tag{2.8}$$

No such precise formula can be given for arbitrary x but a valid estimate is

$$\begin{aligned}|(K^r g)(a)| &\leq \frac{1}{(r-1)!}\int_0^\infty e^{-yt}t^{r-1}[(x-a)^2+t^2]^{\frac{(\lambda-1)}{2}}\,dt \\ &\leq \frac{(x-a)^{\lambda-1}}{(r-1)!}\int_0^\infty e^{-yt}t^{r-1}dt = y^{-r}(x-a)^{\lambda-1}.\end{aligned} \tag{2.9}$$

We are now in a position to prove the next asymptotic result (cf. Lemmas 2.1 and 2.2) for the Fourier transform of a function with singularity.

Theorem 2.10. *Let $\varphi(x)$ be m times continuously differentiable in $[a,b]$. Let φ and its first $m-1$ derivatives vanish at $x = b$. Then, if $0 < \lambda < 1$,*

$$\int_a^b \varphi(x)(x-a)^{\lambda-1}e^{iyx}dx = \sum_{r=0}^{m-1}\frac{\Gamma(r+\lambda)}{r!\,y^{r+\lambda}}e^{iya+i\frac{\pi\lambda}{2}+i\frac{\pi r}{2}}\varphi^{(r)}(a) + O(y^{-m}) \tag{2.11}$$

as $y \to \infty$.

Proof. In the previous notation,

$$\begin{aligned}&\int_a^b \varphi(x)g(x)\,dx \\ &= \left[\sum_{r=0}^{m-1}(-1)^r\varphi^{(r)}(x)(K^{r+1}g)(x)\right]_a^b + (-1)^m\int_a^b \varphi^{(m)}(x)(K^m g)(x)\,dx.\end{aligned}$$

Using (2.8) and estimating, by means of (2.9),

$$\left|\int_a^b \varphi^{(m)}(x)(K^m g)(x)\,dx\right| \le y^{-m}\int_a^b |\varphi^{(m)}(x)|(x-a)^{\lambda-1}\,dx = O(y^{-m}),$$

we complete the proof. □

Remark 2.12. The counterpart of Theorem 2.10 for

$$\int_a^b \varphi(x)(b-x)^{\lambda-1}e^{iyx}dx$$

is completely the same, except that the harmonics in the sum on the left-hand side of (2.11) are $e^{iya-i\frac{\pi\lambda}{2}+i\frac{\pi r}{2}}$ and, of course, in the assumptions and result the roles of a and b are opposite.

A promising extension of Erdélyi's result is given in [214]. Let $D\subset\mathbb{R}^d$ be a domain with compact closure whose boundary $S=\partial D$ may be represented in the form $S=\cup_{j\in J}S_j$, where J is a finite set of indices, and S_j are C^k-smooth hypersurfaces in $\mathbb{R}^d$ in general position. The value $k\ge 1$ will be specified later, but the result remains true in the C^∞ and real analytic setting. The general position assumption implies that the intersections $S_{J'}=\cap_{j\in J'}S_j$, $J'\subseteq J$, if non-void, are C^k-smooth varieties.

We consider functions C^k-smooth inside D whose support is the closure of D, of the form

$$f(x)=\varphi(x)\prod_j \rho_j(x)^{\mu_j}\ln^{m_j}\rho_j(x),$$

where $\varphi(x)\in C^k(\mathbb{R}^d)$, φ vanishes nowhere on ∂D, and

$$\rho_j(x)=\begin{cases}\operatorname{dist}(x,S_j), & \text{if } x\in D,\\ 0, & \text{if } x\notin D.\end{cases} \tag{2.13}$$

As an example $D=\{x\in\mathbb{R}^d : |x|<1\}$ shows, ρ_j may not be smooth functions in the whole of D. Therefore, the product in (2.13) is defined as follows: the factor ρ_j does not appear in it as long as x is far from S_j, so (2.13) must hold locally in D. It is proved in [Gi, Appendix B] that (2.13) indeed defines a C^k-function in a neighborhood of ∂D not including ∂D.

We introduce the following notations:

$$m_+=\sum_{j=1}^a(\mu_j+1),\ m_-=\sum_{j=a+1}^m(\mu_j+1),\ M_+=\sum_{j=1}^a m_j,\quad\text{and}\quad M_-=\sum_{j=a+1}^m m_j.$$

The condition $\mu_j+1>0$, $j\in J$, is assumed throughout.

We give asymptotics of the $\widehat{f}(r\theta)$ as $r \to \infty$ for $\theta \in A$, where A is a subset of the unit sphere $\mathbb{S}^{d-1}$ in $\mathbb{R}^d$ whose complement has Lebesgue measure zero. For $\theta \in A$ we shall denote by $\mathfrak{F}(\theta)$ the set of $x \in \partial D$ such that the hyperplane

$$\left\{x' \in \mathbb{R}^d : \sum_{j=1}^{d} \theta_j(x'_j - x_j) = 0\right\}$$

is not transversal to the largest stratum $S_{J'}$ containing x. Using the partition of unity, we see that the asymptotics is the sum of contributions of $x \in \mathfrak{F}(\theta)$, so it suffices to consider only one contribution. In this case we assume without loss of generality that $J' = \{1, \dots, m\}$.

Theorem 2.14. *The contribution of a point $x \in \mathfrak{F}(\theta)$ to the asymptotics of $\widehat{f}(r\theta)$ as $r \to \infty$ is*

$$r^{-m_+ - m_- - (d-m)/2} \sum_{j=0}^{M_+ + M_-} q_j (\ln r)^{M_+ + M_- - j} \left(e^{ir\theta x} + o(1)\right),$$

where q_j are some numbers.

The coefficient q_0 of the leading term is given explicitly in [Za2] in accordance with different conditions that the numbers m_+, m_-, M_+, M_-, n and m may satisfy.

2.3 The Fourier transform on a convex set

Let us consider a problem of the asymptotic behavior of the Fourier transform of the indicator function of a convex set. First essential results are due to Hlawka [96] in the 40s and 50s. The motivation of his interest in this problem lies in Number Theory, namely, in counting lattice points inside dilates of a convex body. For further progress in estimating the Fourier transform of the indicator, see [94], [161], [162], [137], [185], [204], [36], [37].

A different source of such interest is connected with the Radon transform and important problems related to this notion. Recall that the Radon transform of a function f is defined to be the integral of the restriction of this function to a hyperplane

$$R(f : \theta) = \widetilde{f}(\xi, p) = \int_{\xi x = p} f(x) d\omega, \tag{2.15}$$

where $d\omega$ is the element of the area on the hyperplane $\theta = \{\xi x = p\}$, which is oriented so that it is the boundary of the domain $\xi x < p$.

In [81] the following result is derived from simple geometric observations.

Theorem 2.16. *Let D be a convex symmetric body for which the origin O is the center of symmetry. Let its boundary ∂D be a compact $[\frac{d+3}{2}]$ times differentiable* *hyper-surface in* $\mathbb{R}^d$ *with nowhere-vanishing principal curvatures. Then*

$$\begin{aligned}\widehat{\chi}_D(y) &= \int_D e^{iyx}\,dx \\ &= 2(2\pi)^{\frac{d-1}{2}}(\rho_1\rho_2\cdots\rho_{d-1})^{-\frac{1}{2}} r^{-\frac{d+1}{2}} \cos\left(ar - \frac{d+1}{4}\pi\right)\left(1 + O(r^{-\frac{1}{2}})\right)\end{aligned}$$

provided

$$\left|\cos\left(ar - \frac{d+1}{4}\pi\right)\right| \ge \delta > 0,$$

where $y = r\eta$, $\eta \in \mathbb{S}^{n-1}$, *the unit sphere in* $\mathbb{R}^d$, *a is the distance from O to the two (symmetric) points at which the function* ηx *attains maximum and minimum on* ∂D, *and* $\rho_1, \rho_2, \dots, \rho_{d-1}$ *are the principal curvatures at each of these points.*

To have $o(1)$ in the remainder term instead of the more precise $O(r^{-\frac{1}{2}})$, it suffices to assume the $[\frac{d+2}{2}]$ smoothness.

In the well-known paper [94] by Herz, essentially the same result is obtained under the $[\frac{d+7}{2}]$ smoothness assumption. This paper is motivated by sharp estimates of the remainder term in the asymptotics of the number of lattice points in dilates of D, but this asymptotics of the Fourier transform is frequently used for different goals with extra smoothness assumptions.

A more general result has recently been obtained in [157]. Instead of it we give here its extension from [131] (see also [156]), where more general functions than the indicator are treated.

(A) *We assume throughout this section that D is a convex domain and the principal curvatures of its boundary* $\mathbf{S} = \partial D$ *never vanish.*

Let D have compact closure with ∂D being a C^k smooth hyper-surface, $k \ge 1$.

Let f be a function whose support is the closure of D of the form

$$f(x) = \varphi(x)\rho(x)^\alpha, \tag{2.17}$$

where $\varphi \in C^k(\mathbb{R}^d)$ and does not vanish on ∂D, and $\rho(x)$ is a regularized distance (see, *e.g.*, [179, Ch. VI, Th. 2]) which vanishes outside D, that is, $\rho \in C^\infty$ outside ∂D, $\rho(x) = 0$ for $x \notin D$, and there exist two positive constants C_1 and C_2, $C_1 < C_2$, such that for $x \notin \partial D$,

$$C_1 \operatorname{dist}(x, \partial D) \le \rho(x) \le C_2 \operatorname{dist}(x, \partial D),$$

$\rho(x)$ is C^∞ smooth outside ∂D, and for $\alpha = (\alpha_1, \dots, \alpha_d)$ we have

$$\left|\frac{\partial^\alpha \rho(x)}{\partial x^\alpha}\right| \le C_\alpha[\operatorname{dist}(x, \partial D)]^{1-\alpha_1-\cdots-\alpha_d}.$$

All constants do not depend on ∂D. The point is that the distance itself cannot stand for ρ, since it may be non-smooth for x from ∂D (see, *e.g.*, [86]). These assumptions are close to those preceding Theorem 4.1.

Let $\mathfrak{g}(x) = 0$, $\mathfrak{g} \in C^k$, $\operatorname{grad}\mathfrak{g}(x) \neq 0$, be a local defining equation of $\mathbf{S} = \partial D$. Then (2.17) may be written in the form

$$f(x) = \begin{cases} \mathfrak{g}(x)^\alpha \varphi_0(x), & x \in D, \\ 0, & x \notin D, \end{cases} \tag{2.18}$$

where for $x \in \mathbf{S}$ one has

$$\varphi_0(x) = \varphi(x) \left(\sum_{i=1}^{d} \left(\frac{\partial \mathfrak{g}(x)}{\partial x_i} \right)^2 \right)^{-\frac{\alpha}{2}}. \tag{2.19}$$

The integral (2.15) in the definition of the Radon transform converges if

$$\alpha + 1 > 0. \tag{2.20}$$

Throughout the section condition (2.20) is assumed.

A hyperplane in $\mathbb{R}^d$ may be considered as a point in $\mathbb{RP}_d$, the projective space, so (5.1) defines $R(f;\theta)$ on the whole of $\mathbb{RP}_d$ except for infinite hyper-planes. However, one may extend it continuously to points θ of $\mathbb{RP}_d$ corresponding to infinite hyper-planes by setting $R(f;\theta) = 0$ for these θ (recall that the support of f is compact).

To a smooth hyper-surface $\mathbf{S} \subset \mathbb{R}^d$ (or, more generally, $\mathbf{S} \subset \mathbb{RP}_d$) one may associate the dual variety $\widehat{\mathbf{S}} \subset \mathbb{RP}_d$ defined as the set of hyper-planes not transversal to $\mathbf{S}$ at some point. Recall that the dual of $\widehat{\mathbf{S}}$ is the closure of $\mathbf{S}$ itself, see, *e.g.*, [158, 159] and [213], where further references may be found. Recall that the hyperplane $\theta = \left\{x \in \mathbb{R}^d : \theta_0 + \sum_{i=1}^n \theta_i x_i = 0\right\}$ is not transversal to $\mathbf{S}$ at a point $\bar{x} \in \mathbf{S} \cap \theta$ if

$$(\theta_1 : \cdots : \theta_n) = \left(\frac{\partial \mathfrak{g}(\bar{x})}{\partial x_1} : \cdots : \frac{\partial \mathfrak{g}(\bar{x})}{\partial x_d} \right), \tag{2.21}$$

$\theta_0, \ldots, \theta_d$ are homogeneous coordinates in $\mathbb{RP}_d$.

By [159, Lemma 1], the Radon transform $R(f;\theta)$ is C^k-smooth in $\mathbb{RP}_d \setminus \widehat{\mathbf{S}}$. Condition **(A)** implies that $\widehat{\mathbf{S}}$ is a smooth hyper-surface in $\mathbb{RP}_d$ (see [213] for the discussion of the codimension of the dual variety).

So, let $\bar{\theta} \in \mathbb{RP}_d$ belong to $\widehat{\mathbf{S}}$, and let $\widehat{\mathbf{S}}$ be defined in a neighborhood of $\bar{\theta}$ by an equation $y(\theta) = 0$, $\operatorname{grad} y(\bar{\theta}) \neq 0$. Let again $\bar{x} \in \mathbf{S}$ be the point at which the hyperplane $\bar{\theta}$ is not transversal to $\mathbf{S}$; since D is assumed strictly convex, it follows that the point $\bar{x}$ is uniquely defined. Note that function $y(\theta)$ is defined up to a non-vanishing C^k-factor. Assume $\bar{\theta}_n \neq 0$.

Theorem 2.22. *Let $k > \max(\alpha + \frac{d-1}{2}, 1)$. Then there exists an equation $y(\theta) = 0$ defining $\widehat{\mathbf{S}}$ in a neighborhood U of $\bar{\theta}$, $\operatorname{grad} y \neq 0$ in U, and a function $r(\theta) \in C^k(U)$ such that*

$$R(f;\theta) = y_+^{\alpha + \frac{d-1}{2}} r(\theta). \tag{2.23}$$

If, in a neighborhood of $\bar{x} \in \mathbf{S}$, one has $\mathbf{S} = \{x_d = g(x_1, \dots, x_{d-1})\}$, then one can take $y(\theta) = \left(\theta_d^{-1}\theta_0 - h(-\theta_1\theta_d^{-1}, \dots, -\theta_{d-1}\theta_d^{-1})\right)\epsilon$, where $\epsilon = -1$ if the matrix with the entries $\frac{\partial^2 g}{\partial x_j \partial x_k}$ is positive definite and $\epsilon = 1$ otherwise,

$$h(\beta_1, \dots, \beta_{d-1}) = \sum_{i=1}^{d-1} \beta_i x_i - g(x_1, \dots, x_{d-1}) := Lg,$$

$$\beta_i = \frac{\partial g(x_1, \dots, x_{d-1})}{\partial x_i}, \quad i = 1, \dots, d,$$

Lg is the Legendre transform of $g(x_1, \dots, x_{d-1})$, and

$$r(\bar{\theta}) = \varphi(\bar{x}) \frac{(2\pi)^{\frac{d-1}{2}} |\bar{\theta}_d|^{\alpha + \frac{d-1}{2}} \Gamma(\alpha+1)}{\varkappa^{\frac{1}{2}} \left(\sum_{i=1}^{d} \bar{\theta}_i^2\right)^{\frac{\alpha}{2} + \frac{d-1}{4}} \Gamma(\alpha + \frac{d+1}{2})}. \tag{2.24}$$

Here $\varkappa$ is the Gaussian curvature of $\mathbf{S}$ at the point $\bar{x}$.

Proof. The argument is similar to that in [159]. Without loss of generality we may assume that in a neighborhood of the point $\bar{x}$ the surface $\mathbf{S}$ is given by the equation $x_d = g(x_1, \dots, x_{d-1})$. Thus $x' = (x_1, \dots, x_{d-1})$ may be regarded as local coordinates on $\mathbf{S}$. The statement about the Legendre transform follows from [159] and [213].

It suffices to choose a curve γ in $\mathbb{RP}_d$ which intersects the hyper-surface $\widehat{\mathbf{S}}$ transversally at the point $\bar{\theta}$ and to show that (2.23) holds along this curve. We may choose this curve to be $\gamma = \{\theta : \theta_i = \bar{\theta}_i,\ i = 1, \dots, d\}$, θ_0 is a parameter on γ. Since the dual of $\widehat{\mathbf{S}}$ is $\mathbf{S}$, the hyperplane tangent to $\widehat{\mathbf{S}}$ at the point $\bar{\theta}$ is given by the equation

$$\sum_{i=1}^{d} \bar{x}_i \theta_i + \theta_0 = 0,$$

in which $\bar{x}$ is the point on ∂D defined above. Comparing this equation with the definition of γ we see that γ and the above hyperplane have just one common point, so indeed γ is transversal to $\widehat{\mathbf{S}}$. Since $\bar{\theta}_d \neq 0$, we may consider $\theta_0, \dots, \theta_{d-1}$ as local coordinates in $\mathbb{RP}_d$ near $\bar{\theta}$. Moreover, $R(f;\theta)$ restricted to γ may be regarded as a function of $z = \theta_d^{-1}(\bar{\theta}_0 - \theta_0)$, and we shall write $R(f;z)$ instead of $R(f;\theta)|_\gamma$.

Consider the function $\xi(x) = x_d - g(x_1, \dots, x_{d-1})$. Note that on the hyperplane $\bar{\theta}$ we have

$$\xi = -\bar{\theta}_d^{-1} \left(\sum_{i=1}^{d-1} \bar{\theta}_i x_i + \bar{\theta}_0 \right) - g(x_1, \dots, x_{d-1}).$$

We can use formula (2.18) with $\xi(x)$ instead of $\mathfrak{g}(x)$, then (2.19) yields

$$\varphi_0(\bar{x}) = \varphi(\bar{x})|\bar{\theta}_d|^\alpha \left(\sum_{i=1}^{d} \bar{\theta}_i^2\right)^{-\frac{\alpha}{2}}. \tag{2.25}$$

Note that $(\bar{x}_1, \dots, \bar{x}_{d-1})$ is a critical point of ξ on $\bar{\theta}$. Denote by $\mathfrak{J}$ the determinant of the Hessian of this function at this point, *i.e.*,

$$\mathfrak{J} = \det\left(\left(\frac{-\partial^2 g(\bar{x}_1, \dots, \bar{x}_{d-1})}{\partial x_j \partial x_k}\right)_{j,k=1,\dots,d-1}\right).$$

By assumption **(A)**, we have $\mathfrak{J} \neq 0$; moreover, the Hessian is either a negative or a positive definite quadratic form; in the latter case we may consider $-z$ instead of z as the argument of $R(f;\cdot)$. So we assume that $(-1)^{d-1}\mathfrak{J} > 0$, and by the Morse lemma [144] one may choose coordinates $u_1, \dots, u_{d-1}$ on $\mathbf{S}$ so that for $\theta_0 = \bar{\theta}_0$ we have $\xi = -\sum_{j=1}^{d-1} u_j^2$. Therefore

$$\xi = -\bar{\theta}_d^{-1}(\theta_0 - \bar{\theta}_0) + \xi|_{\theta_0=\bar{\theta}_0} = z - \sum_{j=1}^{d-1} u_j^2.$$

We have

$$\begin{aligned}\left|\det\left(\frac{\partial u_j}{\partial x_k}\right)_{j,k=1,\dots,d-1}\right| &= 2^{\frac{1-d}{2}}((-1)^{d-1}\mathfrak{J})^{\frac{1}{2}} \\ &= 2^{\frac{1-d}{2}}\left|\det\left(\frac{\partial^2 g(\bar{x})}{\partial x_j \partial x_k}\right)_{j,k=1,\dots,d-1}\right|^{\frac{1}{2}}.\end{aligned}$$

Since the Lebesgue measure $d\omega$ on θ is given by

$$d\omega = \frac{\left(\sum\limits_{i=1}^{d} \bar{\theta}_i^2\right)^{\frac{1}{2}}}{|\bar{\theta}_d|} dx_1 \cdots dx_{d-1},$$

one can write (cf. (2.15)):

$$R(f;\theta_0, \bar{\theta}_1, \dots, \bar{\theta}_d) = \int\limits_{\sum_{i=1}^{d-1} u_i^2 \leq z} \left(z - \sum_{i=1}^{d-1} u_i^2\right)^\alpha \varphi_1(u,z) du_1 \cdots du_{d-1},$$

where $z = \bar{\theta}_n^{-1}(\theta_0 - \bar{\theta}_0)$ on θ, and

$$\varphi_1(0,0) = \varphi_0(\bar{x})|\mathfrak{J}|^{-\frac{1}{2}} 2^{\frac{d-1}{2}} \frac{\left(\sum\limits_{i=1}^{d} \bar{\theta}_i^2\right)^{\frac{1}{2}}}{|\bar{\theta}_d|}.$$

Note that the first and the second quadratic forms, G and Q, of $\mathbf{S}$ at the point $\bar{x}$ are given, respectively, by the matrices

$$\left(\delta_{ij} + \frac{\partial g(\bar{x}_1,\ldots,\bar{x}_{d-1})}{\partial x_i}\frac{\partial g(\bar{x}_1,\ldots,\bar{x}_{d-1})}{\partial x_j}\right)_{i,j=1,\ldots,d-1}$$
$$= E + \bar{\theta}_d^{-2}(\bar{\theta}_1,\ldots,\bar{\theta}_{d-1})^t(\bar{\theta}_1,\ldots,\bar{\theta}_{d-1})$$

and

$$\left(1+\sum_{i=1}^{d-1}\left(\frac{\partial g(\bar{x}_1,\ldots,\bar{x}_{d-1})}{\partial x_i}\right)^2\right)^{-\frac{1}{2}}\left(\frac{-\partial^2(\bar{x}_1,\ldots,\bar{x}_{d-1})}{\partial x_i\partial x_j}\right)_{i,j=1,\ldots,d-1}.$$

Here E is the unit $(d-1)\times(d-1)$-matrix. The determinants of these matrices equal, respectively,

$$\frac{1}{\bar{\theta}_d^2}\sum_{i=1}^{d}\bar{\theta}_i^2$$

and

$$|\mathfrak{J}|\left(1+\sum_{i=1}^{d-1}\left(\frac{\partial g(\bar{x}_1,\ldots,\bar{x}_{d-1})}{\partial x_i}\right)^2\right)^{\frac{1-d}{2}} = |\mathfrak{J}|\left(\sum_{i=1}^{d}\bar{\theta}_i^2\right)^{\frac{1-d}{2}}|\bar{\theta}_d|^{d-1}.$$

Therefore, we can write

$$\varphi_1(0,0) = \varphi_0(\bar{x})2^{\frac{d-1}{2}}\varkappa^{\frac{1}{2}}\left(\sum_{i=1}^{d}\bar{\theta}_i^2\right)^{\frac{1-d}{4}}|\bar{\theta}_d|^{\frac{d-1}{2}} \tag{2.26}$$

via the Gaussian curvature $\varkappa$ of $\mathbf{S}$ at the point $\bar{x}$. Thus $R(f;z)=0$ for $z<0$, and for $z>0$ after substituting $u=z^{\frac{1}{2}}t$ we have

$$R(f;z) = z^{\alpha+\frac{d-1}{2}}\int\limits_{\sum_{i=1}^{d-1}t_i^2<1}\varphi_1(tz^{\frac{1}{2}},z)\left(1-\sum_{i=1}^{d-1}t_i^2\right)^{\alpha}dt.$$

Using the spherical coordinates and applying the formula

$$\Omega_d = \frac{2\pi^{\frac{d}{2}}}{\Gamma\left(\frac{d}{2}\right)}$$

for the area Ω_d of the unit sphere in the d-dimensional Euclidean space, we have for $z>0$,

$$R(f;z) = z^{\alpha+\frac{d-1}{2}}\varphi_2(z),$$

where

$$\varphi_2(0) = \varphi_1(0,0)\frac{2\pi^{\frac{d-1}{2}}}{\Gamma\left(\frac{d-1}{2}\right)}\int_0^1 (1-s)^\alpha s^{\frac{d-3}{2}}\,ds.$$

We now use (2.25) and (2.26) and then apply the formula

$$\int_0^1 s^{\mu_1-1}(1-s)^{\mu_2-1}ds = \frac{\Gamma(\mu_1)\Gamma(\mu_2)}{\Gamma(\mu_1+\mu_2)}.$$

The proof of the theorem is complete. □

Remark 2.27. If we do not assume that the principal curvatures of **S** never vanish, then Theorem 2.22 remains valid for almost all directions $(\bar{\theta}_1, \dots, \bar{\theta}_d)$. This follows from the Morse lemma, cf. [159].

As in [157], using the Fourier slice theorem

$$\widehat{f}(r\theta_1, \dots, r\theta_d) = \int_{-\infty}^{\infty} e^{2\pi i r\theta_0} R(f;\theta)\,d\theta_0, \tag{2.28}$$

with $\sum\limits_{i=1}^{d} \theta_i^2 = 1$, and the Erdélyi lemma (see Theorem 2.10 and Remark 2.12), we obtain the asymptotics of the Fourier transform of functions of the type (2.17). This generalizes Theorem 2.15 and its generalizations in [161, 162] and [185]. The result reads as follows.

Theorem 2.29. *Let $\alpha \geq 0$, $k > \max\left(1, \frac{d-1}{2}+\alpha\right)$, $\eta \in \mathbb{R}^d$ be a unit vector, and $x^+(\eta)$ and $x^-(\eta)$ be the (uniquely defined) points of ∂D at which the function ηx attains maximum and minimum on ∂D, respectively. Then for $r \to \infty$,*

$$\widehat{f}(r\eta) = r^{-\alpha-\frac{d+1}{2}}\left[e^{2\pi i r x^+(\eta)\eta}\Xi^+ + e^{2\pi i r x^-(\eta)\eta}\Xi^- + o(1)\right],$$

where

$$\Xi^+ = e^{\pm\pi i\frac{2\alpha+d+1}{4}}\varphi(x^\pm)(\varkappa^\pm)^{-\frac{1}{2}}(2\pi)^{-\alpha-1}\Gamma(\alpha+1),$$

$\varkappa^\pm$ are the Gaussian curvatures of ∂D at the points $x^\pm$, respectively, and the remainder is small uniformly in η.

Observe that the relation (2.28) and the Morse lemma are of crucial importance in these problems. As for applications of Theorem 2.29, besides [131] see also [121] and [130].

For dimension one, the Fourier transform may be converted by the Stationary Phase integral when one substitutes variables in order to simplify the function φ.

In the multivariate case, even if φ is very nice, a phase may appear during an attempt to remove geometrical peculiarities of the domain of integration.

2.4 Applications

The results given in this chapter have various applications. We present some of them, apparently less known.

2.4.1 Hyperbolic means

Let us borrow an example from the estimates of the Lebesgue constants for the so-called hyperbolic linear means (see [125]). The next result needs certain preamble. Since the appearance of Babenko's paper [7] interest has continued in various questions of Approximation Theory and Fourier Analysis connected with the study of linear means with harmonics in the "hyperbolic crosses"

$$\Gamma(N,\gamma) = \left\{k \in \mathbb{Z}^d : h(N,k,\gamma) = \prod_{j=1}^{d} \left(\frac{|k_j|}{N}\right)^{\gamma_j} \le 1,\ \gamma_j > 0,\ j = 1,\dots,d\right\}.$$

We are interested in the hyperbolic means of Bochner–Riesz type of order $\alpha \ge 0$, a hyperbolic analog of the usual (spherical) Bochner–Riesz means,

$$L^{\alpha}_{\Gamma(N,\gamma)} : f(x) \mapsto \sum_{k\in\Gamma(N,\gamma)} (1 - h(N,k,\gamma))^{\alpha}_{+}\, \widehat{f}(k) e^{ikx}.$$

Hyperbolic Bochner–Riesz means (for the two-dimensional Fourier integrals, with $\gamma_1 = \gamma_2 = 2$) first appeared in the paper of El-Kohen [64] in connection with the study of their L^p-norms. His result was not sharp and shortly after was strengthened by Carbery [49].

Starting with the integral

$$\int_{N\mathbb{T}^d} \left| \sum_{k\in\Gamma(N,\gamma)} (1 - h(N,k,\gamma))^{\alpha}_{+}\, e^{2\pi i k\cdot x} \right| dx \tag{2.30}$$

and passing from sums to integrals, we arrive, after appropriate estimates of the errors, at the integral

$$\int_{N\mathbb{T}^d} \left| \int_{\substack{|x_1|^{\gamma_1}\cdots|x_d|^{\gamma_d}\le 1\\ |x_1|,\dots,|x_{d-1}|\ge\frac12}} (1 - |x_1|^{\gamma_1}\cdots|x_d|^{\gamma_d})^{\alpha} e^{iux}\, dx \right| du$$

as the leading term, while the error is proved to be of an appropriate order. Of course, all the other combinations of $d-1$ variables separated from zero should be considered as well, but since they are treated similarly we restrict ourselves to the integral displayed.

Denote the inner integral by $\Psi(u)$; we are interested in the behavior of $\Psi(u)$ for u large. Further, it suffices to consider

$$\Psi(u) = \int\limits_{\substack{x_1^{\gamma_1}\cdots x_d^{\gamma_d}\le 1 \\ x_1,\dots,x_{d-1}\ge\frac12,\ x_d\ge 0}} (1 - x_1^{\gamma_1}\cdots x_d^{\gamma_d})^\alpha \cos(ux)\,dx;$$

keeping the same notation results in no confusion. Introducing a new variable $t = x_1^{\frac{\gamma_1}{\gamma_d}}\cdots x_{d-1}^{\frac{\gamma_{d-1}}{\gamma_d}}\, x_d$, we obtain

$$\Psi(u) = \int_0^1 (1-t^{\gamma_d})^\alpha dt \int_G x_1^{-\frac{\gamma_1}{\gamma_d}}\cdots x_{d-1}^{-\frac{\gamma_{d-1}}{\gamma_d}}$$
$$\times \cos\left(u_1x_1 + \cdots + u_{d-1}x_{d-1} + tu_d x_1^{-\frac{\gamma_1}{\gamma_d}}\cdots x_{d-1}^{-\frac{\gamma_{d-1}}{\gamma_d}}\right) dx_1\cdots dx_{d-1}.$$

Here and in further estimates we denote by the same letter G corresponding domains in $\mathbb{R}_+^{d-1}$. The only essential circumstance here is that the variables are separated from zero and infinity.

Let us change variables once again;

$$v_j = x_j u_j(|u_1|^{\gamma_2}\cdots|u_d|^{\gamma_d}|t|^{\gamma_d})^{-\frac{1}{\gamma}}, \quad j = 1,2,\dots,d-1,$$

with $\gamma = \gamma_1 + \gamma_2 + \cdots + \gamma_d$. We have

$$tu_d x_1^{-\frac{\gamma_1}{\gamma_2}}\cdots x_{d-1}^{-\frac{\gamma_{d-1}}{\gamma_d}} = v_1^{-\frac{\gamma_1}{\gamma_d}}\cdots v_{d-1}^{-\frac{\gamma_{d-1}}{\gamma_d}}(t^{\gamma_d}|u_1|^{\gamma_1}\cdots|u_d|^{\gamma_d})^{\frac{1}{\gamma}}$$

and

$$x_1^{-\frac{\gamma_1}{\gamma_d}}\cdots x_{d-1}^{-\frac{\gamma_{d-1}}{\gamma_d}}\,dx_1\cdots dx_{d-1}$$
$$= t^{\frac{d\gamma_d}{\gamma-1}}(|u_1|^{\gamma_1}\cdots|u_d|^{\gamma_d})^{\frac{d}{\gamma}}|u_1\cdots u_d|^{-1} v_1^{-\frac{\gamma_1}{\gamma_d}}\cdots v_{d-1}^{-\frac{\gamma_{d-1}}{\gamma_d}}\,dv_1\cdots dv_{d-1}.$$

By this we obtain

$$\Psi(u) = |u_1\cdots u_d|^{-1}(|u_1|^{\gamma_1}\cdots|u_d|^{\gamma_d})^{d/\gamma}\int_0^1 (1-t^{\gamma_d})^\alpha\, t^{\frac{d\gamma_d}{\gamma-1}}\,dt$$
$$\int_G \cos\left((t^{\gamma_d}|u_1|^{\gamma_1}\cdots|u_d|^{\gamma_d})^{\frac{1}{\gamma}}\left(v_1+\cdots+v_{d-1}+v_1^{-\frac{\gamma_1}{\gamma_d}}\cdots v_{d-1}^{-\frac{\gamma_{d-1}}{\gamma_d}}\right)\right)$$
$$v_1^{-\frac{\gamma_1}{\gamma_d}}\cdots v_{d-1}^{-\frac{\gamma_{d-1}}{\gamma_d}}\,dv_1\cdots dv_{d-1}.$$

It is convenient to make one more substitution $t^{\frac{\gamma_d}{\gamma}} \to t$. By this we represent Ψ

in the form

$$\Psi(u) = \frac{\gamma}{\gamma_d}|u_1\cdots u_d|^{-1}(|u_1|^{\gamma_1}\cdots|u_d|^{\gamma_d})^{\frac{d}{\gamma}}\int_0^1 (1-t^{\gamma_d})^\alpha t^{d-1}\,dt$$
$$\int_G \cos\left(t(|u_1|^{\gamma_1}\cdots|u_d|^{\gamma_d})^{\frac{1}{\gamma}}\left(v_1+\cdots+v_{d-1}+v_1^{-\frac{\gamma_1}{\gamma_d}}\cdots v_{d-1}^{-\frac{\gamma_{d-1}}{\gamma_d}}\right)\right)$$
$$v_1^{-\frac{\gamma_1}{\gamma_d}}\cdots v_{d-1}^{-\frac{\gamma_{d-1}}{\gamma_d}}\,dv_1\cdots dv_{d-1}.$$

Our next task is to consider the inner integral in the latter representation for Ψ. It can be rewritten to be of the form

$$\int_{\mathbb{R}_+^{d-1}} \varphi(v_1,\ldots,v_{d-1})e^{itM(v_1+\cdots+v_{d-1}+v_1^{-\frac{\gamma_1}{\gamma_d}}\cdots v_{d-1}^{-\frac{\gamma_{d-1}}{\gamma_d}})}\,dv_1\cdots dv_{d-1},$$

where φ is an infinitely differentiable function supported on G, and

$$M = (|u_1|^{\gamma_1}\cdots|u_d|^{\gamma_d})^{\frac{1}{\gamma}}.$$

Writing also $v=(v_1,\ldots,v_{d-1})$ and

$$S(v) = v_1+\cdots+v_{d-1}+v_1^{-\frac{\gamma_1}{\gamma_d}}\cdots v_{d-1}^{-\frac{\gamma_{d-1}}{\gamma_d}},$$

we have to investigate the behavior of the integral

$$Q_n(tM) = \int_{\mathbb{R}_+^{d-1}} \varphi(v)e^{itMS(v)}\,dv.$$

Let us apply Theorem 2.4. We have to find stationary points, if they exist, and calculate all the parameters in (2.5). We obtain for $j=1,2,\ldots,d-1$,

$$\frac{\partial S}{\partial v_j} = 1 - \frac{\gamma_j}{\gamma_d}v_1^{-\frac{\gamma_1}{\gamma_d}}\cdots v_{d-1}^{-\frac{\gamma_{d-1}}{\gamma_d}}v_j^{-1},$$

and solving the system of $d-1$ equations $\frac{\partial S}{\partial v_j}=0$, $j=1,2,\ldots,d-1$, we get a solution

$$v_j^0 = \gamma_j(\gamma_1^{\gamma_1}\cdots\gamma_d^{\gamma_d})^{-\frac{1}{\gamma}}.$$

Let us prove that just this is the unique stationary point. For this we find the value of the determinant of the second derivatives at this point and prove that it is non-zero. We have

$$\frac{\partial^2 S}{\partial v_j\partial v_k} = \begin{cases} \gamma_j\gamma_k\gamma_d^{-2}v_1^{-\frac{\gamma_1}{\gamma_d}}\cdots v_{d-1}^{-\frac{\gamma_{d-1}}{\gamma_d}}v_j^{-1}v_k^{-1}, & j\neq k \\ \frac{\gamma_j}{\gamma_d}\frac{\gamma_j}{\gamma_d+1}v_1^{-\frac{\gamma_1}{\gamma_d}}\cdots v_{d-1}^{-\frac{\gamma_{d-1}}{\gamma_d}}v_j^{-2}, & j=k. \end{cases}$$

This yields

$$\det S''(v) = \gamma_d^{-2(d-1)}(v_1^{-\frac{\gamma_1}{\gamma_d}} \cdots v_{d-1}^{-\frac{\gamma_{d-1}}{\gamma_d}})^{d-1}$$

$$\times \begin{pmatrix} \lambda & \lambda & \lambda & \lambda \\ \frac{\gamma_1(\gamma_1+\gamma_d)}{v_1\, v_1} & \frac{\gamma_1\gamma_2}{v_1\, v_2} & \cdots & \frac{\gamma_1\gamma_{d-1}}{v_1\, v_{d-1}} \\ \frac{\gamma_2\gamma_1}{v_2\, v_1} & \frac{\gamma_2(\gamma_2+\gamma_d)}{v_2\, v_2} & \cdots & \frac{\gamma_2\gamma_{d-1}}{v_2\, v_{d-1}} \\ \cdots & \cdots & \cdots & \cdots \\ \frac{\gamma_{d-1}\gamma_1}{v_{d-1}\, v_1} & \frac{\gamma_{d-1}\gamma_2}{v_{d-1}\, v_2} & \cdots & \frac{\gamma_{d-1}(\gamma_{d-1}+\gamma_d)}{v_{d-1}\, v_{d-1}} \end{pmatrix}$$

$$= \left(v_1^{-\frac{\gamma_1}{\gamma_d}} \cdots v_{d-1}^{-\frac{\gamma_{d-1}}{\gamma_d}}\right)^{d-1} \gamma_d^{-2(d-1)} v_1^{-2} \cdots v_{d-1}^{-2} \gamma_1 \cdots \gamma_{d-1}\, \Delta,$$

where

$$\Delta = \begin{pmatrix} \gamma_1+\gamma_d & \gamma_2 & \cdots & \gamma_{d-1} \\ \gamma_1 & \gamma_2+\gamma_d & \cdots & \gamma_{d-1} \\ \cdots & \cdots & \cdots & \cdots \\ \gamma_1 & \gamma_2 & \cdots & \gamma_{d-1}+\gamma_d \end{pmatrix}.$$

Standard inductive argument yields

$$\Delta = (\gamma_1 + \cdots + \gamma_d)\gamma_d^{d-2} = \gamma\gamma_d^{d-2}.$$

Hence

$$\det S''(v) = \gamma\gamma_1 \cdots \gamma_{d-1}\gamma_d^{-d}(v_1^{-\frac{\gamma_1}{\gamma_d}} \cdots v_{d-1}^{-\frac{\gamma_{d-1}}{\gamma_d}})^{d-1} v_1^{-2} \cdots v_{d-1}^{-2},$$

and

$$\det S''(v_0) = \gamma\gamma_1^{-1} \cdots \gamma_d^{-1}(\gamma_1^{\frac{\gamma_1}{\gamma_d}} \cdots \gamma_{d-1}^{\frac{\gamma_{d-1}}{\gamma_d}}\gamma_d)^{d-1} > 0,$$

therefore v_0 is a stationary point. We now get

$$\begin{aligned} S(v_0) = {} & (\gamma_1^{\gamma_1} \cdots \gamma_d^{\gamma_d})^{-\frac{1}{\gamma}} (\gamma_1 + \cdots + \gamma_{d-1}) \\ & + \left(\gamma_1^{\frac{\gamma_1}{\gamma_d}} \cdots \gamma_{d-1}^{\frac{\gamma_{d-1}}{\gamma_d}}\right)^{-1} (\gamma_1^{\gamma_1} \cdots \gamma_{d-1}^{\gamma_{d-1}}\gamma_d^{\gamma_d})^{\frac{\gamma_1+\cdots+\gamma_{d-1}}{\gamma_d\gamma}} \\ = {} & \gamma\,(\gamma_1^{\gamma_1} \cdots \gamma_d^{\gamma_d})^{-\frac{1}{\gamma}}. \end{aligned}$$

Observe that

$$\begin{aligned} \varphi(v_0) &= (v_1^0)^{-\frac{\gamma_1}{\gamma_d}} \cdots (v_{d-1}^0)^{-\frac{\gamma_{d-1}}{\gamma_d}} \\ &= \gamma_d(\gamma_1^{\gamma_1} \cdots \gamma_d^{\gamma_d})^{-\frac{1}{\gamma}}. \end{aligned}$$

Obviously,

$$\int\limits_{\substack{|u_j|\le 1,\\ j=1,2,\dots,d}} |\Psi(u)|\,du = O(1),$$

and it remains to estimate

$$\int\limits_{\substack{1\le|u_j|\le \pi N\\ j=1,2,\dots,d}} |\Psi(u)|\,du.$$

For the leading term in (2.5), we have

$$\int\limits_{\substack{1\le|u_j|\le \pi N\\ j=1,2,\dots,d}} \frac{(|u_1|^{\gamma_1}\cdots|u_d|^{\gamma_d})^{\frac{d}{\gamma}}}{|u_1\cdots u_d|}(|u_1|^{\gamma_1}\cdots|u_d|^{\gamma_d})^{\frac{d-1}{2\gamma}}$$

$$\times\left|\int_0^1 (1-t^{\gamma_d})^{\alpha} t^{\frac{d-1}{2}} e^{it(|\gamma_1|^{\gamma_1}\cdots|\gamma_d|^{\gamma_d}j)^{-\frac{1}{\gamma}}\Phi(v_0)}\,dt\right|\,du.$$

The inner integral is estimated as above. For $\alpha > \frac{d-1}{2}$, we obtain

$$\int\limits_{\substack{1\le|u_j|\le \pi N\\ j=1,2,\dots,d-1}} |u_1\cdots u_d|^{-1}(|u_1|^{\gamma_1}\cdots|u_d|^{\gamma_d})^{\frac{d}{\gamma}}$$

$$\times(|u_1|^{\gamma_1}\cdots|u_d|^{\gamma_d})^{\frac{d-1}{2\gamma}}(|u_1|^{\gamma_1}\cdots|u_d|^{\gamma_d})^{-\frac{\alpha+1}{\gamma}}\,du$$

$$= 2^d\prod_{j=1}^{d}\int_1^{\pi N} u_j^{-1}u_j^{\frac{\gamma_j}{\gamma}(\frac{d-1}{2}-\alpha)}\,du_j$$

$$= O\Bigg(\prod_{j=1}^{d} N^{\frac{\gamma_j}{\gamma}(\frac{d-1}{2}-\alpha)}\Bigg) = O(N^{\frac{d-1}{2}-\alpha}).$$

We now see how to handle the other terms in (2.5), including the remainder one. It remains to mention that for $\alpha = \frac{d-1}{2}$, the leading term in (2.5) gives the product of d integrals estimated by

$$\omega_d \ln^d N + O(\ln^{d-1} N),$$

and the other terms give better bounds.

2.4.2 Multiple Fourier integrals

Let

$$\int_D \widehat{f}(x)e^{2\pi i u\cdot x}\,dx$$

be the partial Fourier integral defined by a set D. The behavior of partial Fourier integrals with respect to a specifically organized family of such sets characterize approximation properties of f. It is natural to define such a family as a sequence of dilations of a fixed set D. This has been extensively studied when D is the cube (cubic case)

$$D = \{x \in \mathbb{R}^d : |x_j| \le 1,\ j = 1, 2, \dots, d\},$$

or the ball (spherical case)

$$D = \{x \in \mathbb{R}^d : |x| = (x_1^2 + x_2^2 + \cdots + x_d^2)^{\frac{1}{2}} \le 1\}.$$

Their N-dilations are

$$RD = \{x \in \mathbb{R}^d : |x_j| \le N,\ j = 1, 2, \dots, d\}$$

and

$$RD = \{x \in \mathbb{R}^d : |x| \le N\},$$

respectively. The other example of a family of sets is the family of rectangles

$$\{x \in \mathbb{R}^d : |x_j| \le N_j, \quad N_j > 0,\ j = 1, 2, \dots, d\}$$

that cannot be expressed as a family of dilations of a fixed set. Numerous results on these (as well as references) may be found, *e.g.*, in [215, Ch. 17] or [181], where similar problems are studied for multiple Fourier series as well.

Theorem 2.29 is used to obtain weak type estimates in the weighted L^p spaces.

In a general form, the problems of above type were studied in [130]. Let D be a convex domain with nowhere vanishing principal curvatures of its boundary ∂D. Let D have compact closure with ∂D being a C^k-smooth hyper-surface, $k \ge 1$. Set

$$E = E(M, x) = \{u : x - u \in MD\}.$$

Let λ be a function defined as f in (2.17), that is, whose support is the closure of D and that is C^k-smooth inside D, of the form

$$\lambda(x) = \rho(x)^\alpha \varphi(x), \tag{2.31}$$

where $\varphi \in C^k(\mathbb{R}^d)$ and does not vanish on ∂D, and ρ is a regularized distance to the boundary.

Define the linear means of the Fourier integral

$$\sigma_N(f; x) = \sigma_N(f; x; \lambda) = \int_{\mathbb{R}^d} f(x - s) N^d \widehat{\lambda}(-Ns)\, ds. \tag{2.32}$$

Indeed, for f smooth enough we have by Fubini's theorem

$$\begin{aligned}\sigma_N(f;x) &= \int_{\mathbb{R}^d} f(x-s)N^d\widehat{\lambda}(-Ns)\,ds\\ &= \int_{\mathbb{R}^d} f(x-s)N^d\int_{\mathbb{R}^d}\lambda(v)e^{2\pi iNs\cdot v}\,dv\,ds\\ &= \int_{\mathbb{R}^d} f(x-s)\int_{\mathbb{R}^d}\lambda\left(\frac{v}{N}\right)e^{2\pi iv\cdot s}\,dv\,ds\\ &= \int_{\mathbb{R}^d}\lambda\left(\frac{v}{N}\right)e^{2\pi ix\cdot v}\,dv\int_{\mathbb{R}^d} f(s)e^{-2\pi iv\cdot s}\,ds\\ &= \int_{\mathbb{R}^d}\lambda\left(\frac{v}{N}\right)\widehat{f}(v)e^{2\pi ix\cdot v}\,dv,\end{aligned}$$

that is, the linear means are defined in a usual (multiplier) way. We see that they are defined by means of the function λ that, in turn, strongly depends on geometric properties of D. We will write $\sigma_N(f)$ if the argument is of no importance for us. The representation (2.32) is a usual way to avoid problems of definition of the Fourier transform of f.

We are going to restrict ourselves to the critical case $\alpha > \frac{d-1}{2}$. Thus (2.32) becomes a generalization of the Bochner–Riesz means of order greater than the critical one. Also, we fix arbitrary k which satisfies $k > \alpha + \frac{d-1}{2}$.

We present an auxiliary estimate rather than essential results. Just this estimate is quite representative and handy. Set

$$\sigma_*(f;x) = \sup_{R>0} |\sigma_N(f;x)|.$$

Observe that for $\alpha > \frac{d-1}{2}$, the Fourier transform $\widehat{\lambda}$ is integrable on $\mathbb{R}^d$; we have also $k > \alpha + \frac{d-1}{2}$ which justifies the above specification.

Set

$$f_M(x) = \frac{1}{|E(M,x)|}\int_{E(M,x)} |f(u)|\,du,$$

and

$$f^*(x) = \sup_{M>0} f_M(x).$$

Lemma 2.33. *The inequality*

$$\sigma_*(f;x) \le C_{d,\alpha} f^*(x)$$

holds.

Proof. For $N|s|$ small enough it suffices to make use of the inequality

$$\left|\int_{\mathbb{R}^d}\lambda(u)e^{iNsu}du\right| = \left|\int_D \lambda(u)e^{iNsu}du\right| \le C, \tag{2.34}$$

while for $N|s|$ large enough Theorem 2.29 yields

$$\left|\int_{\mathbb{R}^d} \lambda(u)e^{iNsu}du\right| \le C(N|s|)^{-\alpha-\frac{d+1}{2}}. \tag{2.35}$$

We have

$$\begin{aligned}\sigma_N(f;x) = &\int_{E(\frac{2^M}{N},0)} f(x-s)N^d\widehat{\lambda}(-Ns)\,ds\\ &+\sum_{k=M}^{\infty}\int_{\Delta_k(N)} f(x-s)N^d\widehat{\lambda}(-Ns)\,ds\end{aligned} \tag{2.36}$$

where

$$\Delta_k(N) = E\left(\frac{2^{k+1}}{N},0\right)\setminus E\left(\frac{2^k}{N},0\right)$$

and M is such that $x-s\in E(\frac{2^M}{N},0)$. For the first integral on the right-hand side of (2.36), we have, by (2.34),

$$\begin{aligned}&\left|\int_{E(\frac{2^M}{N},0)} f(x-s)\,dsN^d\widehat{\lambda}(-Ns)\right|\\ &\le C\frac{1}{|E\left(\frac{2^M}{N},0\right)|}\int_{E\left(\frac{2^M}{N},0\right)}|f(x-s)|\,ds \le Cf_{\frac{2^M}{N}}(x).\end{aligned} \tag{2.37}$$

Furthermore, by (2.35),

$$\begin{aligned}&\left|\int_{\Delta_k(N)} f(x-s)\,ds\,N^d\widehat{\lambda}(-Ns)\right|\\ &\le N^{\frac{d-1}{2}-\alpha}\int_{\Delta_k(N)}|f(x-s)||s|^{-\alpha-\frac{d+1}{2}}\,ds\\ &\le CN^{\frac{d-1}{2}-\alpha}\left(\frac{2^k}{N}\right)^{-\alpha-\frac{d+1}{2}}\int_{E(\frac{2^{k+1}}{N},0)}|f(x-s)|\,ds\\ &\le CN^d2^{-k\left(\alpha+\frac{d+1}{2}\right)}\int_{E(\frac{2^{k+1}}{N},0)}|f(x-s)|\,ds\\ &\le CN^d2^{-k\left(\alpha+\frac{d+1}{2}\right)}\left(\frac{2^{k+1}}{N}\right)^d\frac{1}{|E\left(\frac{2^{k+1}}{N},0\right)|}\int_{E\left(\frac{2^{k+1}}{N},0\right)}|f(x-s)|\,ds\\ &\le C2^{k\left(\frac{d-1}{2}-\alpha\right)}f_{\frac{2^{k+1}}{N}}(x).\end{aligned} \tag{2.38}$$

It follows from the representation (2.36) and estimates (2.37) and (2.38) that

$$\sigma_*(f,x) \le C\left\{1+\sum_{k=M}^{\infty}2^{k(\frac{d-1}{2}-\alpha)}\right\}f^*(x) \le C_{d,\alpha}f^*(x), \tag{2.39}$$

since the series in (2.39) converges just for $\alpha > \frac{d-1}{2}$. The lemma is proved. □

2.4.3 Generalized Bochner–Riesz means of critical order

The next result not only refines an old classical result of Stein [178] but also shows how precise the results obtained by means of the above estimates can be.

Theorem 2.40. *Let D be the compact support of a function λ with the d-smooth boundary ∂D. Assume that D is convex and the principal curvatures of ∂D never vanish. Then there exists a positive constant $C_{D,\lambda}$ depending only on D and λ such that*

$$\|L_N^\lambda\| = C_{D,\lambda}\ln N + o(\ln N) \tag{2.41}$$

for large N.

The proof relies on two basic facts. The following theorem, which is a part of Belinsky's result [19, Thm. 1], gives the first basic fact.

Let $\Delta_z^m(\lambda; h_1,\dots,h_m)$ be the mth difference of a function λ defined recursively by the formulas

$$\Delta_z^1(\lambda; h_1) = \lambda(z+h_1) - \lambda(z),$$
$$\Delta_z^m(\lambda; h_1,\dots,h_m) = \Delta_{z+h_m}^{m-1}(\lambda; h_1,\dots,h_{m-1}) - \Delta_z^{m-1}(\lambda; h_1,\dots,h_{m-1}),$$

where $h_j, z \in \mathbb{R}$, m is an arbitrary integer.

Theorem A. *For each compactly supported continuous function λ there holds*

$$\begin{aligned}\|L_N^\lambda\| = \int\limits_{N\mathbb{T}^d}\left|\prod_{j=1}^d \frac{\pi x_j}{N\sin\frac{\pi x_j}{N}}\widehat{\lambda}(x)\right| dx + O\Bigg\{\sum_{j=1}^{m-1}\int\limits_{N\mathbb{T}^d}\left|\frac{x}{N}\right|^j |\widehat{\lambda}(x)|\,dx \\ + \int\limits_{\mathbb{T}^d}\cdots\int\limits_{\mathbb{T}^d}\left[\sum_k \left|\Delta_{\frac{k}{N}}^m\left(\lambda; \frac{u_1}{N},\dots,\frac{u_m}{N}\right)\right|^2\right]^{\frac12} du_1\cdots du_m\Bigg\}.\end{aligned} \tag{2.42}$$

The proof of this result may be found also in the Appendix to the paper [131] and in [203, Ch. 9].

In order to apply Theorem A, we need some information about the behavior of the Fourier transform. This is Theorem 2.29, and it is our second basic fact. Moreover, unlike in the previous application, here Theorem 2.29 will be used in its full completeness and sharpness, that is, asymptotics rather than merely upper estimate.

Proof. Let us use (2.42) with m specified below, obtain the asymptotics of the first term in the right-hand side of (2.42), and estimate all the rest from above. Let us start with the estimates of the remainder terms in (2.42). As for the integrals from the sum in the first term of the remainder in (2.42), using spherical coordinates and applying Theorem B, we get

$$\int_0^{CN} N^{-j}t^j t^{d-1} t^{-\frac{d+1}{2}-\frac{d-1}{2}}\,dt \le C_1 N^{-j} N^{\frac{n-1}{2}-\frac{d-1}{2}+j} = O(1) \tag{2.43}$$

for each $j = 1, \ldots, m-1$. Now, consider the last term in (2.42). Our computations are similar to those used by Belinsky for the usual Bochner–Riesz means [19] and to those carried out in the general situation in [131]. It suffices to estimate the bigger quantity

$$\sup_{u_1,\ldots,u_m\in\mathbb{T}} \left\{ \sum_k \left| \Delta^m_{\frac{k}{N}} \left(\lambda; \frac{u_1}{N}, \ldots, \frac{u_m}{N}\right) \right|^2 \right\}^{\frac{1}{2}}$$

$$= \sup_{u_1,\ldots,u_m\in\mathbb{T}} \left\{ \sum_{m+2\le \mathrm{dist}(k,\partial(NS))\le CN} \left| \Delta^m_{\frac{k}{N}} \left(\lambda; \frac{u_1}{N}, \ldots, \frac{u_m}{N}\right) \right|^2 \right. \tag{2.44}$$

$$\left. + \sum_{0\le \mathrm{dist}(k,\partial(NS))\le m+2} \left| \Delta^m_{\frac{k}{N}} \left(\lambda; \frac{u_1}{N}, \ldots, \frac{u_m}{N}\right) \right|^2 \right\}^{\frac{1}{2}}.$$

We may estimate from above each term in the last sum by the maximal value of λ at the points with integer coordinates. This yields

$$\sum_{\substack{0\le \mathrm{dist}(k,\partial(NS))\\ \le m+2}} \left| \Delta^m_{\frac{k}{N}} \left(\lambda; \frac{u_1}{N}, \ldots, \frac{u_m}{N}\right) \right|^2 \le C \sum_{\substack{0\le \mathrm{dist}(k,\partial(NS))\\ \le m+2}} \left(\frac{m+2}{N}\right)^{2\frac{d-1}{2}}.$$

The number of points with integer coordinates inside NY, where $Y \subset \mathbb{R}$ is a convex domain, is equal to $N^d \operatorname{vol}(Y) + O(N^{d-1})$, see, *e.g.*, [204]. Therefore, the last sum is not greater than

$$CN^{-2\frac{d-1}{2}} \sum_{0\le \mathrm{dist}(k,\partial(NS))\le m+2} 1 \le C_1 N^{d-1-2\frac{d-1}{2}} = O(1). \tag{2.45}$$

Now, consider the first sum on the right-hand side of (2.44). Use m times the mean value theorem for the directional derivative. The sum being estimated is transformed to the form

$$N^{-2m} \sum_{m+2\le \mathrm{dist}(k,\partial(NS))\le CN} \left| \frac{\partial^m \lambda\left(\frac{k}{N} + \sigma_1 \frac{u_1}{N} + \cdots + \sigma_m \frac{u_m}{N}\right)}{\partial u_1 \cdots \partial u_m} \right|^2.$$

Here $0 < \sigma_j < 1$, $j = 1, \ldots, m$. We can take any m such that $\frac{d+1}{2} \le m \le d$. If $\frac{d-1}{2}$ is an integer the derivative is bounded, so this quantity is not greater than $CN^{d-2-2\frac{d-1}{2}} = O(1)$. Otherwise, estimating the derivative in the direction $(\sigma_1, \ldots, \sigma_m)$ by its maximal value, we obtain

$$\begin{aligned} N^{-2m} \sum & \left| \frac{\partial^m \lambda\left(\frac{k}{N} + \sigma_1 \frac{u_1}{N} + \cdots + \sigma_m \frac{u_m}{N}\right)}{\partial u_1 \cdots \partial u_m} \right|^2 \\ & \le C_1 N^{-2m} \sum \left[\mathrm{dist}\left(\frac{k+\sigma_{m,k}}{N}, \partial S\right)\right]^{2(\frac{d-1}{2}-m)}, \end{aligned} \tag{2.46}$$

where $0 \le \sigma_{m,k} \le m$ and the sums are over $m+2 \le \operatorname{dist}(k, \partial(NS)) \le CN$. Let us denote $k + \sigma_{m,k}$ by $\bar{k}$. We have provided $m > \frac{d-1}{2}$:

$$\begin{aligned}
&N^{-2m} \sum_{\frac{m+2}{N} \le \operatorname{dist}\left(\frac{k}{N}, \partial S\right) \le C} \left[\operatorname{dist}\left(\frac{\bar{k}}{N}, \partial S\right)\right]^{2\left(\frac{d-1}{2}-m\right)} \\
&= N^{-2m} \sum_{m+2 \le q \le CN} \sum_{\frac{q}{N} \le \operatorname{dist}\left(\frac{k}{N}, \partial S\right) \le \frac{q+1}{N}} \left[\operatorname{dist}\left(\frac{\bar{k}}{N}, \partial S\right)\right]^{d-1-2m} \\
&\le CN^{-2m} \sum_{1 \le q \le CN} \sum_{\frac{q}{N} \le \operatorname{dist}\left(\frac{k}{N}, \partial S\right) \le \frac{q+1}{N}} \left(\frac{q}{N}\right)^{d-1-2m} \\
&= CN^{-n+1} \sum_{1 \le q \le CN} q^{d-1-2m} \sum_{\frac{q}{N} \le \operatorname{dist}\left(\frac{k}{N}, \partial S\right) \le \frac{q+1}{N}} 1 \\
&\le CN^{-d+1} N^{d-1} \sum_{1 \le q \le CN} q^{d-1-2m} \le C(1 + N^{d-1-2m+1}) \\
&\le C(1 + N^{d-2\frac{d+1}{2}}) = O(1).
\end{aligned} \tag{2.47}$$

Collecting (2.43)–(2.47), we see that the remainder in (2.42) is bounded. Now, we go on to an estimate of the leading term. Let us pass to the spherical coordinates and apply Theorem 2.29 with $\alpha = \frac{d-1}{2}$. Taking into account also that

$$\frac{\pi x_j}{N \sin \frac{\pi x_j}{N}} - 1 = O\left(\frac{x_j^2}{N^2}\right),$$

which yields estimates like for the remainder terms (see (2.46)), we get the following in place of the leading term:

$$\int_{|\theta|=1} \int_{\frac{1}{2\pi}}^{\frac{N}{2}} \left| \Xi^+(\theta) e^{2\pi i t x^+(\theta)\theta} + \Xi^-(\theta) e^{2\pi i t x^-(\theta)\theta} \right| \frac{dt}{t} + o(\ln N).$$

Consider the inner integral in an equivalent form:

$$\int_1^{\pi N} \left| \Xi^+(\theta) e^{it(x^+(\theta) - x^-(\theta))\theta} + \Xi^-(\theta) \right| \frac{dt}{t},$$

and estimate it, denoting $(x^+(\theta) - x^-(\theta))\theta$ by $d(\theta)$. The following relation is very well known (see, *e.g.*, [215, Vol. 1, Ch. 2]):

$$\int_1^{\pi N} \left| \frac{\sin t}{t} \right| dt = \frac{2}{\pi} \ln N + O(1). \tag{2.48}$$

Let us adopt the method used in [215] to get (2.48) with our conditions. The estimated integral is

$$\int_{d(\theta)}^{\pi Nd(\theta)} \left|\Xi^+(\theta)e^{it} + \Xi^-(\theta)\right| \frac{dt}{t} = \sum_{k=1}^{[Nd(\theta)]} \int_{2k\pi}^{2(k+1)\pi} \left|\Xi^+(\theta)e^{it} + \Xi^-(\theta)\right| \frac{dt}{t} + O(1).$$

The integral on the right-hand side is equal to

$$\begin{aligned}\sum_{k=1}^{[Nd(\theta)]} &\int_{\frac{2k\pi}{Nd(\theta)}}^{\frac{2(k+1)\pi}{Nd(\theta)}} \left|\Xi^+(\theta)e^{itNd(\theta)} + \Xi^-(\theta)\right| \frac{dt}{t}\\ &= \int_0^{\frac{2\pi}{Nd(\theta)}} \left|\Xi^+(\theta)e^{itNd(\theta)} + \Xi^-(\theta)\right| \sum_{k=1}^{[Nd(\theta)]} \frac{1}{t + \frac{2k\pi}{Nd(\theta)}}\, dt.\end{aligned}$$

The last sum is equal to $\frac{Nd(\theta)}{2\pi}[\ln Nd(\theta) + O(1)]$. Thus, we need to estimate

$$\begin{aligned}\frac{1}{2\pi}&(2\pi)^{-\frac{d+1}{2}}\Gamma\left(\frac{d+1}{2}\right)\int_{|\theta|=1} Nd(\theta)[\ln Nd(\theta)\\ &\quad + O(1)]\,d\theta \int_0^{\frac{2\pi}{Nd(\theta)}} \left|\Xi^+(\theta)e^{itNd(\theta)} + \Xi^-(\theta)\right| dt\\ &= (2\pi)^{-\frac{d+3}{2}}\Gamma\left(\frac{d+1}{2}\right)\int_{|\theta|=1} [\ln Nd(\theta) + O(1)]\,d\theta \int_0^{2\pi} \left|\Xi^+(\theta)e^{it} + \Xi^-(\theta)\right| dt\\ &= \ln N (2\pi)^{-\frac{d+3}{2}}\Gamma\left(\frac{d+1}{2}\right)\int_{|\theta|=1} d\theta \int_0^{2\pi} \left|\Xi^+(\theta)e^{it} + \Xi^-(\theta)\right| dt + O(1).\end{aligned}$$

Denoting $f(x^\pm(\theta))(\varkappa^\pm(\theta))^{-\frac{1}{2}}$ by $\phi^\pm(\theta)$ (it should be recalled that f and the Gaussian curvature $\varkappa$ do not vanish anywhere), we arrive at

$$\begin{aligned}(2\pi)^{-\frac{d+3}{2}}&\Gamma\left(\frac{d+1}{2}\right)\int_{|\theta|=1} d\theta \int_0^{2\pi} |\phi^+(\theta)e^{\frac{i\pi d}{2}}e^{it} + \phi^-(\theta)e^{-\frac{i\pi d}{2}}|\,dt\\ &= (2\pi)^{-\frac{d+3}{2}}\Gamma\left(\frac{d+1}{2}\right)\int_{|\theta|=1} d\theta \int_0^{2\pi} |(-1)^d\phi^+(\theta)e^{it} + \phi^-(\theta)|\,dt.\end{aligned} \tag{2.49}$$

But the expression in the inner integral on the right-hand side of (2.49) never vanishes, and the right-hand side of (2.49) may be denoted by $C_{S,\lambda}$. The proof is complete. □

Remark 2.50. Let us find $C_{S,\lambda}$ for usual Bochner–Riesz means of critical order, in (2.41). Here S is the unit sphere, so $\varkappa = 1$ everywhere. Since $\lambda(x) = (1-|x|^2)_+^{\frac{d-1}{2}}$, the function $f(x) = (1+|x|)^{\frac{d-1}{2}}$. On the boundary, that is for $|x| = 1$, we get $f(x^{\pm}(\theta)) = 2^{\frac{d-1}{2}}$. Taking into account the following well-known identity

$$\int_{|\theta|=1} d\theta = \frac{2\pi^{\frac{d}{2}}}{\Gamma\left(\frac{d}{2}\right)},$$

we get

$$\begin{aligned} C_{S,\lambda} &= (2\pi)^{-\frac{d+3}{2}}\Gamma\left(\frac{d+1}{2}\right)\frac{2\pi^{\frac{d}{2}}}{\Gamma\left(\frac{d}{2}\right)}2^{\frac{d-1}{2}}\int_0^{2\pi}|(-1)^d e^{it}+1|\,dt \\ &= \frac{1}{2}\pi^{-\frac{3}{2}}\frac{\Gamma(\frac{d+1}{2})}{\Gamma\left(\frac{d}{2}\right)}8 = 4\Gamma\left(\frac{d+1}{2}\right)\Big/\pi^{\frac{3}{2}}\Gamma\left(\frac{d}{2}\right). \end{aligned}$$

The same value was obtained in [22] and, after correcting a small misprint, in [19].

Chapter 3

The Fourier Transform of Convex and Oscillating Functions

What may be referred to as an initial point for the subject of this chapter is Trigub's result of the 1970's on the asymptotics of the Fourier transform of a convex function (see, *e.g.*, [198]). Roughly speaking, the Fourier transform of a convex function can be represented as a handy leading term and an integrable function. This was a generalization and strengthening of Shilov's result [172] on the asymptotics for the Fourier coefficients of a convex function (see also [10, Ch. IX, §6]). We shall present extensions of Trigub's result for functions from much wider classes. But prior to this we will consider a compromise case when the function itself is convex but multiplied by certain singularities; one endpoint or two at both of the endpoints is enough for complete understanding of the situation.

3.1 Convex functions with singularities

We do not formulate Trigub's result just now (Corollary 3.15 or more general Corollary 3.16 below), since it is a partial case of the following more general theorem (see [119]). We add singularities at the endpoints of the interval where a convex function is defined, in this case a sort of asymptotic formula can readily be obtained by uniting Erdélyi's and Trigub's techniques.

Theorem 3.1. *Let φ be convex on $[a,b]$ and $2d = \min\{b-a,1\}$. Then for any $|y|\ge 1$, $0<\lambda\le 1$, $0<\mu\le 1$, we have*

$$\int_a^b (x-a)^{\lambda-1}(b-x)^{\mu-1}\varphi(x)e^{2\pi iyx}\,dx$$
$$= \Gamma(\lambda)(2\pi y)^{-\lambda}\varphi\left(a+\frac{d}{|y|}\right)e^{2\pi iya+\frac{\lambda\pi i}{2}} + \Gamma(\mu)(2\pi y)^{-\mu}\varphi\left(b-\frac{d}{|y|}\right)e^{2\pi iyb-\frac{\mu\pi i}{2}}$$
$$+O\left[\int_a^{a+\frac{2d}{|y|}}|\varphi'(x)|(x-a)^{\lambda}dx+\int_{b-\frac{2d}{|y|}}^b|\varphi'(x)|(b-x)^{\mu}dx\right]. \tag{3.2}$$

Proof. The result is a generalization of Corollary 3.16; the latter can be obtained by taking $\lambda = \mu = 1$. Instead of (3.2), let us prove the following relation, which in a similar manner extends Corollary 3.15. Recall that in the previous chapter we introduced notations

$$g(x) = e^{2\pi iyx}(b-x)^{\mu-1},$$

and for $r = 1, 2, \dots,$

$$(K^r g)(x) = \frac{1}{(r-1)!}\int_{-\infty i}^{x} (x-t)^{r-1} g(t)dt,$$

with integration along the ray in the complex plane going out from x to infinity parallel to the imaginary axis in the negative direction.

Lemma 3.3. *Let φ be convex on $[a,b]$ with $|\varphi'(a)| < \infty$. Then for $|y| \geq 1, 0 < \mu < 1$, we have*

$$\begin{aligned}
&\int_a^b \varphi(x)(b-x)^{\mu-1}e^{2\pi iyx}dx \\
&\quad = \Gamma(\mu)(2\pi y)^{-\mu}\varphi\left(b-\frac{d}{|y|}\right)e^{2\pi iyb-\frac{\mu\pi i}{2}} \\
&\quad - \varphi(a)(K^1 g)(a) + \Theta_1\left[\frac{|\varphi'(a)|d^{\mu-1}}{y^{\mu+1}} + \frac{(b-a)^{\mu-1}}{y^2}\right] \\
&\quad + \Theta_2\int_{b-\frac{d}{|y|}}^{b} |\varphi'(x)|(b-x)^{\mu}\, dx,
\end{aligned} \tag{3.4}$$

where $|\Theta_1| \leq (2\pi)^{-2}$ and $|\Theta_2| \leq \frac{1}{\mu} + \frac{2^{\mu+2}}{(2\pi d)^2(2^{\mu+1}-1)}$.

Proof. First, we make the following transformation:

$$\begin{aligned}
&\int_a^b \varphi(x)(b-x)^{\mu-1}e^{2\pi iyx}dx \\
&\quad = \varphi(b-\frac{d}{|y|})\int_a^b (b-x)^{\mu-1}e^{2\pi iyx}dx \\
&\quad + \int_a^b \left[\varphi(x) - \varphi\left(b-\frac{d}{|y|}\right)\right](b-x)^{\mu-1}e^{2\pi iyx}dx \\
&\quad = \varphi\left(b-\frac{d}{|y|}\right)\int_a^b (b-x)^{\mu-1}e^{2\pi iyx}dx \\
&\quad + \int_a^{b-\frac{d}{|y|}} \left[\varphi(x) - \varphi\left(b-\frac{d}{|y|}\right)\right](b-x)^{\mu-1}e^{2\pi iyx}dx \\
&\quad + \int_{b-\frac{d}{|y|}}^{b} \left[\varphi(x) - \varphi\left(b-\frac{d}{|y|}\right)\right](b-x)^{\mu-1}e^{2\pi iyx}dx.
\end{aligned} \tag{3.5}$$

Recall that in Section 2 of the previous chapter (see, *e.g.*, also [66], [57, §11]) it was shown that (cf. Remark 2.12)

$$(K^r g)(b) = e^{2\pi i y b - \frac{\mu\pi i}{2} - \frac{(r+1)\pi i}{2}} \frac{\Gamma(r+\mu-1)}{(r-1)!} (2\pi y)^{1-r-\mu} \tag{3.6}$$

and

$$(K^r g)(x) \le (b-x)^{\mu-1}(2\pi y)^{-r}. \tag{3.7}$$

Using (3.7), we obtain

$$\begin{aligned}
&\varphi\left(b - \frac{d}{|y|}\right) \int_a^b (b-x)^{\mu-1} e^{2\pi i y x} dx \\
&= \varphi\left(b - \frac{d}{|y|}\right) [(K^1 g)(b) - (K^1 g)(a)] \\
&= \varphi\left(b - \frac{d}{|y|}\right) \left[\Gamma(\mu) e^{2\pi i y b - \frac{\mu\pi i}{2}} (2\pi y)^{-\mu} - (K^1 g)(a)\right].
\end{aligned} \tag{3.8}$$

Further,

$$\begin{aligned}
&\left| \int_{b-\frac{d}{|y|}}^b \left[\varphi(x) - \varphi\left(b - \frac{d}{|y|}\right)\right] (b-x)^{\mu-1} e^{2\pi i y x} dx \right| \\
&= \left| \int_{b-\frac{d}{|y|}}^b (b-x)^{\mu-1} e^{2\pi i y x} dx \int_{b-\frac{d}{|y|}}^x \varphi'(t)\, dt \right| \\
&= \left| \int_{b-\frac{d}{|y|}}^b \varphi'(t)\, dt \int_t^b (b-x)^{\mu-1} e^{2\pi i y x} dx \right| \\
&\le \mu^{-1} \int_{b-\frac{d}{|y|}}^b |\varphi'(t)| (b-t)^\mu dt.
\end{aligned} \tag{3.9}$$

Let us estimate the remaining integral. Integrating by parts twice, we obtain

$$\begin{aligned}
&\int_a^{b-\frac{d}{|y|}} \left[\varphi(x) - \varphi\left(b - \frac{d}{|y|}\right)\right] (b-x)^{\mu-1} e^{2\pi i y x} dx \\
&= -\left[\varphi(a) - \varphi\left(b - \frac{d}{|y|}\right)\right] (K^1 g)(a) - \int_a^{b-\frac{d}{|y|}} \varphi'(x)(K^1 g)(x) dx \\
&= -\left[\varphi(a) - \varphi\left(b - \frac{d}{|y|}\right)\right] (K^1 g)(a) - \varphi'(x)(K^2 g)(x) \Big|_a^{b-\frac{d}{|y|}} \\
&\quad + \int_a^{b-\frac{d}{|y|}} (K^2 g)(x) d\varphi'(x).
\end{aligned} \tag{3.10}$$

Here, φ' may be understood as a one-sided derivative of φ. Since φ is convex, φ'

is monotone. This and (3.7) yield

$$\begin{aligned}\left|\int_a^{b-\frac{d}{|y|}}(K^2g)(x)d\varphi'(x)\right| &\le (2\pi y)^{-2}\int_a^{b-\frac{d}{|y|}}(b-x)^{\mu-1}|d\varphi'(x)|\\ &\le (2\pi)^{-2}d^{\mu-1}|y|^{-\mu-1}\left|\int_a^{b-\frac{d}{|y|}}d\varphi'(x)\right| \qquad (3.11)\\ &\le (2\pi)^{-2}d^{\mu-1}|y|^{-\mu-1}\left|\varphi\left(-\frac{d}{|y|}\right)\right|\\ &\quad+(2\pi)^{-2}d^{\mu-1}|y|^{-\mu-1}|\varphi'(a)|.\end{aligned}$$

Besides, it follows from (3.7) that

$$\left|\varphi'\left(b-\frac{d}{|y|}\right)(K^2g)\left(b-\frac{d}{|y|}\right)\right| \le \left|\varphi'\left(b-\frac{d}{|y|}\right)\right| d^{\mu-1}(2\pi|y|)^{-\mu-1}, \qquad (3.12)$$

and

$$|\varphi'(a)(K^2g)(a)| \le |\varphi'(a)|(b-a)^{\mu-1}(2\pi y)^{-2}. \qquad (3.13)$$

Again, in view of the convexity of φ, we have

$$\begin{aligned}&(2\pi)^{-2}d^{\mu-1}|y|^{-\mu-1}\left|\varphi'\left(b-\frac{d}{|y|}\right)\right|\\ &\le \frac{(\mu+1)2^{\mu+1}d^{\mu-1}}{4\pi^2d^{\mu+1}(2^{\mu+1}-1)}\int_{b-2\frac{d}{|y|}}^{b-\frac{d}{2|y|}}|\varphi'(t)|(b-t)^{\mu}dt \qquad (3.14)\\ &\le \frac{\mu+1}{(2\pi d)^2}\,\frac{2^{\mu+1}}{2^{\mu+1}-1}\int_{b-2\frac{d}{|y|}}^{b}|\varphi'(t)|(b-t)^{\mu}dt.\end{aligned}$$

To get (3.4), it remains to combine (3.5) and (3.8)–(3.14). The lemma is proved. □

The counterpart of this lemma with singularity at a is obvious. To complete the proof of the theorem, it remains, for some $c\in(a,b)$, to apply Lemma 3.3 and its counterpart to the integrals over $[c,b]$ and $[a,c]$, respectively. □

Denoting the remainder term in (3.2) by $\gamma(y)$, we should notice the following features of this function. First, γ is monotone decreasing as $|y|\to\infty$; this might be important in applications. Secondly, for $\lambda=\mu=1$, the function γ is integrable and

$$\int_1^{\infty}|\gamma(y)|\,dy \le C\int_a^b|\varphi'(x)|dx.$$

We obtain Trigub's result mentioned above by letting $\lambda=\mu=1$; this leads to the following asymptotic formulas for the Fourier transform of the convex function φ on $[a,b]$ with no singularities of Erdélyi type. The first corollary is of less generality, with additional constraints at one of the endpoints.

Corollary 3.15. *If f is convex on $[a,b]$, where $-\infty < a < b \le +\infty$, and $|f'(b)| < \infty$, then for each $r \in \mathbb{R}$, $|r| \ge 1$,*

$$\int_a^b f(t)\, e^{-2\pi irt}\, dt = \frac{i}{r} \left\{ f(b)\, e^{-2\pi ibr} - f\left(a + \frac{d}{|r|}\right) e^{-2\pi iar} \right\} + \theta\gamma(|r|),$$

where $2d = \min\{b-a, 1\}$, $|\theta| \le C$, and γ is monotone decreasing so that

$$\int_1^\infty |\gamma(t)|\, dt \le \frac{1}{d} V_f + |f'(b)|.$$

The other corollary is of full generality as compared with the previous one, but actually is an almost immediate consequence of it.

Corollary 3.16. *If f is convex on $[a,b]$, where $-\infty < a < b \le +\infty$, then for each $r \in \mathbb{R}$, $|r| \ge 1$,*

$$\int_a^b f(t)\, e^{-2\pi irt}\, dt = \frac{i}{r} \left\{ f(b - \frac{d}{|r|})\, e^{-2\pi ibr} - f\left(a + \frac{d}{|r|}\right) e^{-2\pi iar} \right\} + \theta\gamma(|r|),$$

where $2d = \min\{b-a, 1\}$, $|\theta| \le C$, and γ is monotone decreasing so that

$$\int_1^\infty |\gamma(t)|\, dt \le \frac{1}{d} V_f.$$

Here V_f denotes the bounded variation of f.

3.2 Asymptotic behavior in a wider sense

Every classical book on asymptotic expansions (see, *e.g.*, [66]) starts with explanations of the key word "asymptotic" in various contexts. We have special reasons to follow this tradition. Let us discuss how we understand formulas that represent asymptotic behavior of the Fourier transform. Surely, we have real asymptotics in the formulas Stationary Phase Method gives, in the sense of classical analysis when two functions f and g are asymptotically equivalent in the neighborhood of the point x_0 if

$$\lim_{x \to x_0} \frac{f(x)}{g(x)} = 1.$$

In an equivalent form this reads as

$$f(x) = g(x) + \text{ remainder term(s)}. \tag{3.17}$$

This form delivers an additional field of activity: for example, the decay of the remainder term is frequently of interest and importance. When the left-hand side of (3.17) is the Fourier transform of some function, we will call such formulas the asymptotic representation of the Fourier transform in a wider sense. More precisely, the remainder terms may be treated with respect to some other property rather than only better decay at infinity; of course, they are of special interest when the leading term(s) does not satisfy that property. For instance, belonging to the space of integrable functions like in (3.2) for $\lambda = \mu = 1$ (see the corollaries next to it) or to some other space may be such a property. For all its debatableness, we use the word *asymptotics* in this not very precise meaning and hope that this will result in no confusion.

For "good" functions, that is, smooth enough functions, (3.2) is not asymptotic in the classical sense, moreover does not make much sense in any other sense, except maybe some simple upper estimates, but if a function φ is "bad" enough, say of low Hölder smoothness or logarithmic, the leading terms on the right-hand side of (3.2) are not integrable, and (3.2) turns out to be of a real (asymptotic!) value. Clearly, formulas like (3.2) are elaborated just for "bad" functions; if the function is too smooth, one may arrive at the integral with a function of lower smoothness by means of integration by parts. This is just the way to apply (3.2) to Bessel functions and derive the known asymptotic formulas for them – first integrate the well-known integral representation for the Bessel function by parts to make the integral "bad" enough, and then use (3.2) to obtain a purely asymptotic formula.

3.3 Integrability of trigonometric series

The next problem is whether asymptotic formulas of above type may be obtained for the Fourier transform of a function from a wider class. A simple observation that the right-hand side of (3.2) is linear yields its validity for the class of quasi-convex functions QC, that is, those representable by the difference of two convex functions.

More subtle results have their source in the theory of integrability of trigonometric series. The main question of this theory reads as follows.

Given a trigonometric series

$$\frac{a_0}{2} + \sum_{k=1}^{\infty}(a_k \cos 2\pi kx + b_k \sin 2\pi kx), \tag{3.18}$$

find assumptions on the sequences of coefficients $\{a_k\}, \{b_k\}$ under which the series is the Fourier series of an integrable function. We will say in this case that the trigonometric series is integrable (not quite correct but brief and understandable term), or that the sequence belongs to $\widehat{L^1}$.

Frequently, the series

$$\frac{a_0}{2} + \sum_{k=1}^{\infty} a_k \cos 2\pi kx \tag{3.19}$$

and

$$\sum_{k=1}^{\infty} b_k \sin 2\pi kx \tag{3.20}$$

are investigated separately, since there is a difference in their behavior. Usually, integrability of (3.20) requires additional assumptions.

To the best of our knowledge, there exists no convenient description of $\widehat{L^1}$ in terms of a given sequence alone. Actually, there are some characterizations, *e.g.*, [66, 164, 165, 166], but they are too complicated to be applied to concrete problems and they involve properties of functions. Hence, subspaces of $\widehat{L^1}$ are studied so that they are both as wide as possible and described in terms convenient for applications.

First of all, in view of the Riemann–Lebesgue lemma $\widehat{L^1}$ itself is a subspace of c_0, the space of null sequences.

In 1922, Sidon [173] (see also [10, Vol. I]) gave an example of an even monotone null sequence which is not in $\widehat{L^1}$. This means that also the space of sequences of bounded variation

$$bv = \left\{d = \{d_k\} : \|d\|_{bv} = \sum_{k=0}^{\infty} |\Delta d_k| < \infty\right\}$$

is not a subspace of $\widehat{L^1}$. Here $\Delta d_k = d_k - d_{k+1}$.

Let us give some examples of spaces χ which being subspaces of bv are also subspaces of $\widehat{L^1}$. There are many others of course but those proved to be of the most interest and importance.

1) The so-called Boas–Telyakovskii space (see [188]–[191]). Let

$$s_d = \sum_{n=2}^{\infty} \left| \sum_{k=1}^{[\frac{n}{2}]} \frac{\Delta d_{n-k} - \Delta d_{n+k}}{k} \right|, \tag{3.21}$$

then

$$bt = \{d = \{d_k\} : \|d\|_{bt} = \|d\|_{bv} + s_d < \infty\}.$$

This was Telyakovskii's generalization of Boas' result in the way that in [27] the sign of absolute value in (3.21) was inside the second sum. In fact, (3.21) first appeared in [111], where a different though related problem was considered.

2) Fomin's space [74] (cf. [35]):

$$a_p = \{d = \left\{ d_k\} : \|d\|_{a_p} = \sum_{n=0}^{\infty} 2^{\frac{n}{p'}} \left\{ \sum_{k=2^n}^{2^{n+1}-1} |\Delta d_k|^p \right\}^{\frac{1}{p}} < \infty \right\},$$

where $1 < p < \infty$ and $\frac{1}{p} + \frac{1}{p'} = 1$.

3) The Sidon–Telyakovskii conditions [192]:

$$A_k \downarrow 0 \ (k \to \infty), \quad \sum_{k=0}^{\infty} A_k < \infty, \quad \text{and} \quad |\Delta d_k| < A_k.$$

Equivalently, these conditions define the sequence $\{d_k\}$ to belong to the space a_∞.

First of all, we are interested in **1)** and its extensions, since the spaces defined in **2)** and **3)** are subspaces of bt.

A typical strong result due to Telyakovskii is the following

Theorem 3.22. *Let $\{a_k\}$, $\{b_k\}$ be null sequences. Then*

$$\int_0^{\frac{1}{2}} \left| \frac{a_0}{2} + \sum_{k=1}^{\infty} a_k \cos 2\pi kx \right| dx = O(\|a\|_{bv} + s_a),$$

and uniformly with respect to $p = 1, 2, \dots$

$$\int_{\frac{1}{p+\frac{1}{2}}}^{\frac{1}{2}} \left| \sum_{k=1}^{\infty} b_k \sin 2\pi kx \right| dx = 2\pi \sum_{k=1}^{p} \frac{|b_k|}{k} + O(\|b\|_{bv} + s_b),$$

and trigonometric series (3.19) *and* (3.20) *are the Fourier series.*

Analyzing the part of Theorem 3.22 about sine series, one sees that the "endpoints" of the sequence, its beginning and infinity, play a special role. In the papers by Bausov [13] and Telyakovskii [191] results were given in which an arbitrary member of a sequence was allowed to play such a special role. The following result of Telyakovskii is the most general form of this type of results.

Let

$$s_d^m = \sum_{n=2}^{\infty} \left| \sum_{k=1}^{[\frac{n}{2}]} \frac{\Delta d_{m+n-k} - \Delta d_{m+n-k}}{k} \right|,$$

obviously $s_d^0 = s_d$. Denote

$$q_{n,m} = \min \left(\left[\frac{n}{2} \right], \left[\frac{m-n}{2} \right] \right)$$

and

$$\xi_k = \xi \left(b_k, \sqrt{(a_{m-k} - a_{m+k})^2 + (b_{m-k} - b_{m+k})^2} \right)$$

with

$$\xi(t,u) = \begin{cases} \frac{\pi|t|}{2}, & |u| \le |t| \\ |t| \arcsin|\frac{t}{u}| + \sqrt{u^2 - t^2}, & |t| < |u|. \end{cases}$$

What is really essential for ξ is that (see [191, Lemma 5])

$$\xi(t,u) \le \frac{\pi|t|}{2} + |u|, \quad \xi(t,u) \ge \frac{\pi|t|}{2}, \quad \xi(t,u) \ge |u|,$$

and (see, *e.g.*, [191, Lemma 4])

$$\pi \int_{-\frac{1}{2}}^{\frac{1}{2}} |A + B\sin 2\pi x + D\cos 2\pi x|\, dx = 2\xi(A, \sqrt{B^2 + D^2}). \tag{3.23}$$

Theorem 3.24. *Let $\{a_k\}$ and $\{b_k\}$ in (3.18) be null sequences. Then there holds*

$$\left| \int_{-\frac{1}{2}}^{\frac{1}{2}} \left| \frac{a_0}{2} + \sum_{k=1}^{\infty} a_k \cos 2\pi kx + b_k \sin 2\pi kx \right| dx - 4\left\{ 2\sum_{k=1}^{m} \frac{\xi_k}{k} + \pi \sum_{k=2m+1}^{\infty} \frac{|b_k|}{k} \right\} \right|$$
$$\le C\Bigg(\|a\|_{bv} + s_a^m + \|b\|_{bv} + s_b^m + \sum_{n=2}^{m-2} \left| \sum_{k=1}^{q_{n,m}} \frac{\Delta a_{n-k} - \Delta a_{n+k}}{k} \right|$$
$$+ \sum_{n=2}^{m-2} \left| \sum_{k=1}^{q_{n,m}} \frac{\Delta b_{n-k} - \Delta b_{n+k}}{k} \right| \Bigg)$$

uniformly with respect to $m = 0, 1, 2, \ldots$.

One of the strongest known conditions that ensures (along with certain other natural assumptions) the integrability of trigonometric series, pulled in [6] and [47], can be described as follows. Let the space of sequences $\{d_n\}$ be endowed with the norm

$$\|\{d_n\}\|_{a_{1,2}} = \sum_{m=0}^{\infty} \left\{ \sum_{j=1}^{\infty} \left[\sum_{n=j2^m}^{(j+1)2^m - 1} |d_n| \right]^2 \right\}^{\frac{1}{2}} < \infty. \tag{3.25}$$

It is of amalgam nature; the reader can consult on the theory of various amalgam spaces in [72], [75], [92].

It is proved in [6] and [47] that if the coefficients $\{a_n\}$ in (3.19) tend to 0 as $n \to \infty$ and the sequence $\{\Delta a_n\}$ is in $a_{1,2}$, then (3.19) represents an integrable function on $[0, \frac{1}{2}]$. In parallel, if $\{\Delta b_n\} \in a_{1,2}$, then (3.20) represents an integrable function on $[0, \frac{1}{2}]$ if and only if

$$\sum_{n=1}^{\infty} \frac{|b_n|}{n} < \infty. \tag{3.26}$$

It is worth mentioning that $a_{1,2} \subset \ell^1$, and, correspondingly, $\{\Delta a_n\} \in a_{1,2}$ and $\{\Delta b_n\} \in a_{1,2}$ mean that both are bv sequences.

3.4 Analogous function spaces

We will consider function analogs of the spaces of sequences from the previous section for getting pointwise asymptotic behavior of the Fourier transform of such functions rather than just integrability as we had for series. If a result is valid for all types of such spaces, or no matter which of these spaces is used, we will denote it by X instead of specific notation.

For $1 < q < \infty$, set

$$\|g\|_X := \|g\|_{A_q} = \int_0^\infty \left(\frac{1}{u} \int_{u \le |t| \le 2u} |g(t)|^q dt \right)^{\frac{1}{q}} du.$$

These spaces and their sequence analogs (see **2**) above) first appeared in the paper by D. Borwein [35], but became – for sequences – widely known after the paper by Fomin [74]; see also [82, 83]. Later on, these spaces appeared as a partial case of so-called Herz spaces (see first of all the initial paper by Herz [95], note also a relevant paper of Flett [73]).

Further, for $q = \infty$ let

$$\|g\|_X := \|g\|_{A_\infty} = \int_0^\infty \operatorname*{ess\,sup}_{u \le |t| \le 2u} |g(t)| du.$$

The role of an integrable monotone majorant for problems of almost everywhere convergence of singular integrals is known from the work of D.K. Faddeev (see, *e.g.*, [2, Ch. IV, §4]; also [181, Ch. I]); for spectral synthesis problems it was used by Beurling [26], for more details see [24].

Finally, let

$$\|g\|_X := \|g\|_{H_{BT}} = \int_{\mathbb{R}} |g(t)|\, dt + \int_{\mathbb{R}} \left| \int_0^{u/2} \frac{g(u-t) - g(u+t)}{t} \right| du. \tag{3.27}$$

This space was first introduced in [119] as a generalization of **1**). Recently, in a paper by Fridli [77], the inner integral in the last summand on the right of (3.27) was called the Telyakovskii transform. This makes sense and really deserves a separate study, first of all in connection with the real Hardy space, as we will see below. Among various equivalent definitions of the real Hardy space we shall use the one that defines it as the subspace of integrable functions with integrable Hilbert transform.

Lemma 3.28. *The following embeddings hold:*

$$A_\infty \hookrightarrow A_{p_1} \hookrightarrow A_{p_2} \hookrightarrow H_{BT} \hookrightarrow L^1 \; (p_1 > p_2 > 1). \tag{3.29}$$

Proof. The first two embeddings are merely the results of applying Hölder's inequality. Indeed, with $\frac{p_2}{p_1} > 1$,

$$\int_0^\infty \left(\frac{1}{u}\int_{u\le|t|\le 2u} |g(t)|^{p_1}\,dt\right)^{\frac{1}{p_1}} du$$

$$\le \int_0^\infty \left(\frac{1}{u}\left(\int_{u\le|t|\le 2u} |g(t)|^{p_2}\,dt\right)^{\frac{p_1}{p_2}}\left(\int_{u\le|t|\le 2u} dt\right)^{1-\frac{p_1}{p_2}}\right)^{\frac{1}{p_1}} du$$

$$= 2^{\frac{1}{p_1}-\frac{1}{p_2}}\int_0^\infty \left(u^{-\frac{p_2}{p_1}+(1-\frac{p_1}{p_2})\frac{p_2}{p_1}}\int_{u\le|t|\le 2u} |g(t)|^{p_2}\,dt\right)^{\frac{1}{p_2}} du$$

$$= 2^{\frac{1}{p_1}-\frac{1}{p_2}}\int_0^\infty \left(\frac{1}{u}\int_{u\le|t|\le 2u} |g(t)|^{p_2}\,dt\right)^{\frac{1}{p_2}} du,$$

and we are done; the instance $q = \infty$, the left embedding, obviously goes along the same lines.

To prove the embedding of A_q into L^1, only a standard expedient that we will permanently use is needed:

$$2\ln 2\int_0^\infty |g(t)|\,dt = \int_0^\infty \frac{1}{u}\int_{u\le|t|\le 2u} |g(t)|\,dt\,du.$$

Applying then Hölder's inequality to the inner integral on the right yields

$$\int_0^\infty u^{-1}\int_{u\le|t|\le 2u} |g(t)|\,dt\,du$$

$$\le \int_0^\infty \frac{1}{u}\left(\int_{u\le|t|\le 2u} |g(t)|^q\,dt\right)^{\frac{1}{q}}\left(\int_{u\le|t|\le 2u} dt\right)^{1-\frac{1}{q}} du$$

$$= 2^{1-\frac{1}{q}}\int_0^\infty \left(\frac{1}{u}\int_{u\le|t|\le 2u} |g(t)|^q dt\right)^{\frac{1}{q}} du.$$

A bit more delicate are estimates for the second term on the right-hand side of (3.27). First, because of Lemma 3.40, we can deal with

$$\ln 3\int_0^\infty \left|\int_{\frac{x}{2}}^{\frac{3x}{2}} \frac{g(t)}{x-t}\,dt\right| dx = \int_0^\infty \frac{1}{u}\int_{\frac{u}{2}}^{\frac{3u}{2}}\left|\int_{\frac{x}{2}}^{\frac{3x}{2}} \frac{g(t)}{x-t}\,dt\right| dx\,du.$$

The following estimates are similar to those that have been done in [74] but mainly reduces to M. Riesz's theorem. Indeed, the last integral does not exceed

$$\int_0^\infty \frac{1}{u}\int_{\frac{u}{2}}^{\frac{3u}{2}}\left\{\left|\int_{\frac{u}{2}}^{\frac{x}{2}} \frac{g(t)}{x-t}\,dt\right| + \left|\int_{\frac{3x}{2}}^{\frac{3u}{2}} \frac{g(t)}{x-t}\,dt\right| + \left|\int_{\frac{u}{2}}^{\frac{3u}{2}} \frac{g(t)}{x-t}\,dt\right|\right\} dx\,du.$$

The estimates are the same for the first and the second integrals in braces.

For example,

$$
\begin{aligned}
&\int_0^\infty \frac{1}{u} \int_{\frac{u}{2}}^{\frac{3u}{2}} \left| \int_{\frac{u}{2}}^{\frac{x}{2}} \frac{g(t)}{x-t}\, dt \right| dx\, du \\
&\le \int_0^\infty \frac{1}{u} \int_{\frac{u}{2}}^{\frac{3u}{2}} \left\{ \int_{\frac{u}{2}}^{\frac{x}{2}} |g(t)|^q\, dt \right\}^{\frac{1}{q}} \left\{ \int_{\frac{u}{2}}^{\frac{x}{2}} x - t^{\frac{q}{1-q}}\, dt \right\}^{1-\frac{1}{q}} dx\, du \\
&\le 2^{\frac{1}{q}} \int_0^\infty \frac{1}{u} \int_{\frac{u}{2}}^{\frac{3u}{2}} \left\{ \int_{\frac{u}{2}}^{\frac{x}{2}} |g(t)|^q\, dt \right\}^{\frac{1}{q}} \left(\frac{x}{2}\right)^{(1-\frac{q}{q-1})(1-\frac{1}{q})} dx\, du \\
&= 2^{\frac{2}{q}} \int_0^\infty \frac{1}{u} \int_{\frac{u}{2}}^{\frac{3u}{2}} \left\{ \frac{1}{x} \int_{\frac{u}{2}}^{\frac{x}{2}} |g(t)|^q\, dt \right\}^{\frac{1}{q}} dx\, du \\
&\le 2^{\frac{2}{q}} \int_0^\infty \frac{1}{u} \int_{\frac{u}{2}}^{\frac{3u}{2}} \left\{ \frac{1}{x} \int_{\frac{x}{3}}^{\frac{x}{2}} |g(t)|^q\, dt \right\}^{\frac{1}{q}} dx\, du \\
&= 2^{\frac{2}{q}} \ln 3 \int_0^\infty \left\{ \frac{1}{x} \int_{\frac{x}{3}}^{\frac{x}{2}} |g(t)|^q\, dt \right\}^{\frac{1}{q}} dx.
\end{aligned}
$$

It now remains to make use of the following simple relation (see [119]): for any real numbers α and β, $0 < \alpha < \beta$,

$$
\begin{aligned}
\alpha^{1-\frac{1}{q}} \int_0^\infty \left(\frac{1}{u} \int_{\alpha u}^{\beta u} |g(t)|^q dt \right)^{\frac{1}{q}} du &\le \int_0^\infty \left(\frac{1}{u} \int_u^\infty |g(t)|^q dt \right)^{\frac{1}{q}} du \\
&\le \frac{1}{\alpha^{\frac{1}{q}-1} - \beta^{\frac{1}{q}-1}} \int_0^\infty \left(\frac{1}{u} \int_{\alpha u}^{\beta u} |g(t)|^q dt \right)^{\frac{1}{q}} du.
\end{aligned}
\tag{3.30}
$$

For the last summand in braces, applying again the Hölder inequality yields

$$
\int_0^\infty \frac{1}{u} \int_{\frac{u}{2}}^{\frac{3u}{2}} \left| \int_{u/2}^{\frac{3u}{2}} \frac{g(t)}{x-t}\, dt \right| dx\, du \le \int_0^\infty \frac{1}{u} \left\{ \int_{-\infty}^\infty |\widetilde{G}(t)|^q\, dt \right\}^{\frac{1}{q}} \left\{ \int_{\frac{u}{2}}^{\frac{3u}{2}} dt \right\}^{1-\frac{1}{q}} du,
\tag{3.31}
$$

where $\widetilde{G}$ is the Hilbert transform, up to a constant, of the function G which is equal to g on $[\frac{u}{2}, \frac{3u}{2}]$ and vanishes otherwise. By the M. Riesz theorem,

$$
\left\{ \int_{-\infty}^\infty |\widetilde{G}(t)|^q\, dt \right\}^{\frac{1}{q}} \le C_q \left\{ \int_{-\infty}^\infty |G(t)|^q\, dt \right\}^{\frac{1}{q}} = C_q \left\{ \int_{\frac{u}{2}}^{\frac{3u}{2}} |g(t)|^q\, dt \right\}^{\frac{1}{q}},
$$

and the right-hand side of (3.31) is estimated as follows:

$$
\le C_q \int_0^\infty \frac{1}{u} \left\{ \int_{\frac{u}{2}}^{\frac{3u}{2}} |g(t)|^q\, dt \right\}^{\frac{1}{q}} \left\{ \int_{\frac{u}{2}}^{\frac{3u}{2}} dt \right\}^{1-\frac{1}{q}} du = C_q \int_0^\infty \left\{ \frac{1}{u} \int_{\frac{u}{2}}^{\frac{3u}{2}} |g(t)|^q\, dt \right\}^{\frac{1}{q}} du,
$$

and applying (3.30) completes the proof of the lemma. □

For examples on the difference (proper embedding) between these spaces, see [126].

Let us introduce a function space $A_{1,2}$ as an analog of (3.25). We say that a locally integrable function g defined on $\mathbf{R}_+$ belongs to $A_{1,2}$ if

$$\|g\|_{A_{1,2}} = \sum_{m=-\infty}^{\infty} \left\{ \sum_{j=1}^{\infty} \left[\int_{j2^m}^{(j+1)2^m} |g(t)|\, dt \right]^2 \right\}^{\frac{1}{2}} dx < \infty. \tag{3.32}$$

This space is of amalgam nature as well, since each of the summands in m is the norm in the Wiener amalgam space $W(L^1, \ell^2)$ for functions $2^m g(2^m t)$, where ℓ^p, $1 \le p < \infty$, is a space of sequences $\{d_j\}$ endowed with the norm

$$\|\{d_j\}\|_{\ell^p} = \left(\sum_{j=1}^{\infty} |d_j|^p \right)^{\frac{1}{p}}$$

and the norm of a function $g : \mathbf{R}_+ \to \mathbf{C}$ from the amalgam space $W(L^1, \ell^2)$ is taken as

$$\left\| \left\{ \int_j^{j+1} |g(t)|\, dt \right\} \right\|_{\ell^2}.$$

In other words, we can rewrite (3.32) as follows:

$$\|g\|_{A_{1,2}} = \sum_{m=-\infty}^{\infty} \|2^m g(2^m \cdot)\|_{W(L^1, \ell^2)} < \infty.$$

Like for trigonometric series, where the results are given in terms of belonging of the summable sequences $\{\Delta a_n\}$, $\{\Delta b_n\}$ to $a_{1,2}$, it is similarly expected that new conditions for the integrability of the cosine and sine Fourier transforms will be given in terms of belonging of the derivative of the considered function to $A_{1,2}$. This is possible only if $A_{1,2}$ is a subspace of L^1. Indeed, this follows from

$$\|g\|_{A_{1,2}} \ge \sum_{m=-\infty}^{\infty} \int_{2^m}^{2^{m+1}} |g(t)|\, dt = \|g\|_{L^1(\mathbf{R}_+)}. \tag{3.33}$$

For our aims, this can be reformulated as follows: if $f' \in A_{1,2}$, then f is of bounded variation, that is, $f' \in L^1(\mathbf{R}_+)$.

Before proceeding to estimates of the Fourier transforms, let us compare $A_{1,2}$ with H_{BT}. The point is that the classes H_{BT} and $A_{1,2}$ are incomparable. Counterexamples for sequences can be found in [6] and [76]. Without this fact, the effectiveness of the amalgam type results could be doubtful, both for sequences and functions.

3.5 Asymptotics of the Fourier transform

For functions with derivatives in each of these spaces, either $X = A_q$, $1 < q \le \infty$, or $X = H_{BT}$, or $A_{1,2}$, analogs of the results on the integrability of trigonometric series can be obtained.

3.5.1 Hardy type spaces

The following result is obtained in [119] (for A_∞ it was earlier obtained by Trigub [201]).

Theorem 3.34. *Let f be a locally absolutely continuous function on $\mathbb{R} \setminus \{0\}$, and let $\lim\limits_{|t|\to\infty} f(t) = 0$ and $f' \in X$. Then for $|y| > 0$,*

$$\hat{f}(y) = \frac{i}{2\pi y}\left(f\left(\frac{1}{4|y|}\right) - f\left(\frac{1}{4|y|}\right)\right) + \gamma(y), \tag{3.35}$$

where $\int_{\mathbb{R}} |\gamma(y)|dy \le C\|f'\|_X$.

Since a function from QC and all the more a convex function is that with derivative in A_∞, this is an extension of previous results, say Corollary 3.16. Indeed, QC functions on $(0, \infty)$ are those satisfying

$$\int_0^\infty x|g''(x)|\,dx.$$

And for a locally absolutely continuous function f vanishing at infinity we have

$$\int_0^\infty \operatorname*{ess\,sup}_{u\le t\le 2u} |f'(t)|\,du \le \int_0^\infty \int_u^\infty |f''(x)|\,dx\,du = \int_0^\infty x|f''(x)|\,dx.$$

In view of (3.29), Theorem 3.34 is of the widest generality for $X = BT$; for other X in (3.29) embedded in H_{BT} the relation (3.35) follows immediately. Nevertheless, though H_{BT} is wider than the other spaces X, the conditions of belonging to smaller spaces are easier to verify.

In fact, all these spaces are subspaces of the space of functions of bounded variation. As for the definition of bounded variation, we are not going to concentrate on various details. On the contrary, following Bochner [31, Ch. 1], we will mainly restrict ourselves to functions with Lebesgue integrable derivative, since every such function is equivalent to a function of bounded variation in the sense that it is representable as a linear combination (generally, with complex coefficients) of monotone functions. Without loss of generality, it suffices to prove this for real-valued functions. Indeed, let f have integrable derivative in $[a, b]$. Then it is representable as

$$\begin{aligned} f(x) &= f(b) + \int_x^b \frac{|f'(t)| - f'(t)}{2}\,dt - \int_x^b \frac{|f'(t)| + f'(t)}{2}\,dt \\ &= f(b) + h_1(x) - h_2(x). \end{aligned} \tag{3.36}$$

Both functions $h_1(x)$ and $h_2(x)$ are monotone decreasing. If $b = \infty$, we just consider $\lim_{x\to\infty} f(x)$. Since in that case

$$\lim_{x\to\infty} h_1(x) = \lim_{x\to\infty} h_2(x) = 0,$$

a function of bounded variation vanishing at infinity can be represented as a difference of two monotone decreasing functions, each of them tending to zero at infinity. Of course, the usual definition that applies to the uniform boundedness of the sums of oscillations of a function over all possible systems of non-overlapping intervals might be helpful.

The space H_{BT} is of importance not only because of (3.29) but also because of its proximity to the real Hardy space $H := H(\mathbb{R})$, the space of functions $g \in L^1(\mathbb{R})$ for which their Hilbert transform belongs to $L^1(\mathbb{R})$ as well. The Hilbert transform $\tilde{g}$ of an integrable function g is defined in the principal value sense as

$$\tilde{g}(x) = \frac{1}{\pi}\int_{\mathbb{R}} \frac{g(t)}{x-t}\,dt = \lim_{\varepsilon\to 0}\frac{1}{\pi}\int_{|x-t|\ge\varepsilon>0} \frac{g(t)}{x-t}\,dt.$$

A different way to define the Hilbert transform is

$$\tilde{g}(x) = \lim_{\varepsilon\to 0}\frac{1}{\pi}\int_{\mathbb{R}} \frac{g(t)(x-t)}{(x-t)^2+\varepsilon^2}\,dt.$$

In many cases these two definitions are equivalent, but sometimes either one is more convenient for concrete applications. It is obvious that the inner integral in the second summand on the right-hand side of (3.27) is very close to $\tilde{g}$. More precisely, (3.27) means (cf. Lemma 3.40) not that g itself belongs to $H(\mathbb{R})$, but that each of its parts on the positive and on the negative half-axis being extended in the odd way to the second half-axis, respectively, does belong to $H(\mathbb{R})$. By the way, this does not mean that merely the odd part of g belongs to $H(\mathbb{R})$. For recent study of these connections, see [77]. We mention that the remainder term in Theorem 3.34 is not monotone, unlike that in Theorem 3.1 and in the corresponding corollaries.

For the sake of convenience, we will prove the result in a somewhat different form.

Let us denote by φ the odd extension of f' from $[0,\infty)$ to the whole $\mathbb{R}$.

Theorem 3.37. *Let f be a locally absolutely continuous function on $[0,\infty)$, let $\lim_{t\to\infty} f(t) = 0$, and $\varphi \in H$. Then for $y > 0$*

$$F_c(y) = \int_0^\infty f(x)\cos 2\pi yx\,dx = \gamma_1(y), \tag{3.38}$$

and

$$F_s(y) = \int_0^\infty f(x)\sin 2\pi yx\,dx = \frac{1}{2\pi y} f\left(\frac{1}{4y}\right) + \gamma_2(y), \tag{3.39}$$

where $\int_0^\infty |\gamma_j(y)|dy \le C\|\varphi\|_H$, $j = 1, 2$.

The possibility to prove the theorem in this form is justified by the following assertion.

Lemma 3.40. *Let g be an odd function integrable on $\mathbb{R}$. Then*

$$\int_{\mathbb{R}} \frac{g(t)}{x-t}\,dt = \int_{\frac{x}{2}}^{\frac{3x}{2}} \frac{g(t)}{x-t}\,dt + \gamma(x),$$

where

$$\int_{\mathbb{R}} |\gamma(x)|\,dx \le C \int_{\mathbb{R}} |g(t)|\,dt.$$

Proof. We may assume, without loss of generality, that $x > 0$ since for $x < 0$ the proof is exactly the same. Substituting $t \to -t$ in the integral

$$I_1 = \int_{-\infty}^{-\frac{3x}{2}} \frac{g(t)}{x-t}\,dt,$$

we obtain

$$I_1 = \int_{\frac{3x}{2}}^{\infty} \frac{g(-t)}{x+t}\,dt = -\int_{\frac{3x}{2}}^{\infty} \frac{g(t)}{x+t}\,dt.$$

We have

$$\int_{\frac{3x}{2}}^{\infty} \frac{g(t)}{x-t}\,dt + I_1 = \int_{\frac{3x}{2}}^{\infty} g(t)\left[\frac{1}{x-t} - \frac{1}{x+t}\right] dt,$$

and by Fubini's theorem

$$\begin{aligned}\int_0^\infty \left|\int_{\frac{3x}{2}}^{\infty} g(t)\left[\frac{1}{x-t} - \frac{1}{x+t}\right] dt\right| dx &\le \int_0^\infty |g(t)|\,dt \int_0^{\frac{2t}{3}} \left[\frac{1}{x+t} - \frac{1}{x-t}\right] dx \\ = \int_0^\infty |g(t)|\,dt \ln\frac{x+t}{t-x}\Big|_0^{\frac{2t}{3}} &= \int_0^\infty |g(t)| \ln\frac{\frac{2t}{3}+t}{\frac{t}{3}}\,dt \\ &= \ln 5 \int_0^\infty |g(t)|\,dt.\end{aligned}$$

In the same way

$$\int_{-\frac{x}{2}}^{0} \frac{g(t)}{x-t}\,dt = \int_0^{\frac{x}{2}} \frac{g(-t)}{x+t}\,dt = -\int_0^{\frac{x}{2}} \frac{g(t)}{x+t}\,dt,$$

and

$$\begin{aligned}\int_0^\infty \left|\int_0^{\frac{x}{2}} g(t)\left[\frac{1}{x-t} - \frac{1}{x+t}\right] dt\right| dx &\le \int_0^\infty |g(t)| \int_{2t}^\infty \left[\frac{1}{x-t} - \frac{1}{x+t}\right] dx\,dt \\ &= \int_0^\infty |g(t)|\,dt \ln\frac{x-t}{t+x}\Big|_{2t}^\infty = \ln 3 \int_0^\infty |g(t)|\,dt,\end{aligned}$$

which completes the proof. $\square$

Remark 3.41. For g odd, we have the so-called odd Hilbert transform

$$\frac{1}{\pi}\int_{\mathbb{R}}\frac{g(t)}{x-t}\,dt=\frac{1}{\pi}\int_0^\infty g(t)\left[\frac{1}{x-t}-\frac{1}{x+t}\right]dt=\frac{2}{\pi}\int_0^\infty\frac{tg(t)}{x^2-t^2}\,dt. \tag{3.42}$$

We are now in a position to prove Theorem 3.37.

Proof. Integrating by parts yields

$$\begin{aligned}F_s(y)&=\frac{1}{2\pi y}f(0)+\frac{1}{2\pi y}\int_0^{\frac{1}{4y}}f'(x)\,dx+\frac{1}{2\pi y}\int_0^{\frac{1}{4y}}f'(x)(\cos 2\pi yx-1)\,dx\\&\quad+\frac{1}{2\pi y}\int_{\frac{1}{4y}}^\infty f'(x)\cos 2\pi yx\,dx\\&=\frac{1}{2\pi y}f\left(\frac{1}{4y}\right)+O\left(\int_0^{\frac{1}{4y}}|f'(x)|x\,dx\right)+\frac{1}{2\pi y}\int_{\frac{1}{4y}}^\infty f'(x)\cos 2\pi yx\,dx.\end{aligned}$$

Similarly,

$$F_c(y)=O\left(\int_0^{\frac{1}{4y}}|f'(x)|x\,dx\right)-\frac{1}{2\pi y}\int_{\frac{1}{4y}}^\infty f'(x)\sin 2\pi yx\,dx.$$

Since

$$4\int_0^\infty dy\int_0^{\frac{1}{4y}}|f'(x)|x\,dx=\int_0^\infty|f'(x)|\,dx,$$

it remains to show that the last integral in both representations, for F_s and F_c, satisfies the assumptions imposed on γ_j. We will prove this in detail for F_s, since for F_c computations are similar.

Thus we examine, for sufficiently large N,

$$\int_0^N\left|\int_{\frac{1}{4y}}^\infty f'(x)\cos 2\pi yx\,dx\right|\frac{dy}{y}.$$

Writing

$$\Phi(y)=\begin{cases}\displaystyle\int_{\frac{1}{4y}}^\infty f'(x)\cos 2\pi yx\,dx, & 0\le y\le N,\\ \displaystyle\left(2-\frac{y}{N}\right)\int_{\frac{1}{4N}}^\infty f'(x)\cos 2\pi yx\,dx, & N<y\le 2N,\\ 0, & y>2N,\end{cases}$$

we have

$$\int_0^N\left|\int_{\frac{1}{4y}}^\infty f'(x)\cos 2\pi yx\,dx\right|\frac{dy}{y}\le\int_0^\infty\frac{|\Phi(y)|}{y}\,dy.$$

The continuation of the proof may seem strange, in a sense. To estimate the Fourier transform of the initial function, we estimate the Fourier transform of Φ, roughly speaking the Fourier transform of the Fourier transform. This is reflected by the next result which is a generalization of a result by Boas in [27]. For functions from $H(\mathbb{R})$ such a result is just an extension of the Hardy–Littlewood theorem (see, *e.g.*, [215]).

Lemma 3.43. *Let* $\Phi \in L^1(0,\infty)$. *Then*

$$\int_0^\infty \frac{|\Phi(y)|}{y}\,dy \le C\int_0^\infty |\hat{\Phi}(u)|\,du. \tag{3.44}$$

Proof. The following short proof is contained in essence in [25].We have

$$\begin{aligned} \frac{1}{R}\sum_{\alpha R\le k\le \beta R}\left|\Phi\left(\frac{k}{R}\right)\right|R/k &\le \sum_{k=1}^{\infty}\left|\Phi\left(\frac{k}{R}\right)\right|k \\ &\le C\int_{-1/2}^{1/2}\left|\sum_{k=1}^{\infty}\Phi\left(\frac{k}{R}\right)e^{2\pi ikx}\right|dx \le C\int_0^\infty\left|\hat{\Phi}(u)\right|\,du. \end{aligned} \tag{3.45}$$

For sufficiently large R and an appropriate choice of α and β, the first inequality is obvious, the second one is the Hardy–Littlewood theorem, and the last one is well known (see, *e.g.*, [20], where it is given under much more general assumptions). We observe that an integral sum for the left-hand side of (3.44) occurs on the left of (3.45). Passing then to the limit as $R \to \infty$ completes the proof. $\square$

We now return to the proof of the theorem. We are going to estimate the Fourier transform of Φ. Since

$$e^{2\pi iuy} = \cos 2\pi uy + i\sin 2\pi uy,$$

we restrict ourselves to estimating, say, the sine Fourier transform of Φ; the cosine Fourier transform is estimated in the same way with minor changes. We have

$$\begin{aligned} \int_0^\infty \Phi(y)\sin 2\pi uy\,dy &= \int_0^N \sin 2\pi uy\int_{\frac{1}{4y}}^\infty f'(x)\cos 2\pi yx\,dx\,dy \\ &\quad + \int_N^{2N}\left(2-\frac{y}{N}\right)\sin 2\pi uy\int_{\frac{1}{4N}}^\infty f'(x)\cos 2\pi yx\,dx\,dy. \end{aligned}$$

Changing the order of integration, we arrive at the integral

$$\int_{\frac{1}{4N}}^\infty f'(x)\left[\int_{\frac{\pi}{2x}}^N \sin 2\pi uy\cos 2\pi yx\,dy + \int_N^{2N}\left(2-\frac{y}{N}\right)\sin 2\pi uy\cos 2\pi yx\,dy\right]dx.$$

Using known trigonometric formulas for both inner integrals and integrating by parts in the second one, we get

$$\begin{aligned}
&\frac{1}{4\pi}\int_{\frac{1}{4N}}^{\infty} f'(x)\left\{\left[-\frac{\cos 2\pi(u+x)y}{u+x}-\frac{\cos 2\pi(u-x)y}{u-x}\right]_{\frac{1}{4x}}^{N}\right.\\
&\quad+\left(2-\frac{y}{N}\right)\left[-\frac{\cos 2\pi(u+x)y}{u+x}-\frac{\cos 2\pi(u-x)y}{u-x}\right]_{N}^{2N}\\
&\quad\left.+\frac{1}{N}\int_{N}^{2N}\left[-\frac{\cos 2\pi(u+x)y}{u+x}-\frac{\cos 2\pi(u-x)y}{u-x}\right]dy\right\}dx \qquad (3.46)\\
&=\frac{1}{4\pi}\int_{\frac{1}{4N}}^{\infty} f'(x)\left[\frac{1}{u+x}-\frac{1}{u-x}\right]\sin\frac{\pi u}{2x}\,dx\\
&\quad+\frac{1}{4\pi N}\int_{\frac{1}{4N}}^{\infty} f'(x)\int_{N}^{2N}\left[\frac{\cos 2\pi(u+x)y}{u+x}+\frac{\cos 2\pi(u-x)y}{u-x}\right]dy\,dx.
\end{aligned}$$

We are now going, in accordance with Lemma 3.43, to estimate different parts of the right-hand side with respect to integrability in u. Grouping these parts in a special way, we then integrate them modulo over $[\frac{1}{2N},\infty)$. Indeed, integration over $[0,\frac{1}{2N}]$ is carried out trivially:

$$\begin{aligned}
\int_0^{\frac{1}{2N}}|\hat{\Phi}(u)|\,du &\le \int_0^{\frac{1}{2N}}\left[\left(\int_0^{N}+\int_N^{2N}\right)dy\int_0^\infty|f'(x)|\,dx\right]du\\
&=\int_0^\infty|f'(x)|\,dx.
\end{aligned}$$

We mostly deal with the terms corresponding to $u-x$ since those corresponding to $u+x$ are handled even easier. First, grouping

$$\int_{\frac{1}{2N}}^{\infty} f'(x)\,\frac{1}{N}\int_N^{2N}\frac{1}{u-x}\left[\cos 2\pi(u-x)y-\sin\frac{\pi u}{2x}\right]dy\,dx$$

and estimating the expression in the square brackets by

$$2\left|\sin\left(\frac{u-x}{2}\left(\frac{\pi}{2x}-2\pi y\right)\right)\right|,$$

we proceed to the part when

$$x\in\left[u-\frac{1}{4N},u+\frac{1}{4N}\right]$$

as follows:

$$\begin{aligned}
&\int_{\frac{1}{2N}}^{\infty}\left|\int_{u-\frac{1}{4N}}^{u+\frac{1}{4N}} f'(x)\,\frac{1}{N}\int_{N}^{2N}\left(2\pi y-\frac{\pi}{2x}\right)dy\,dx\right|du\\
&\quad\le \frac{1}{N}\int_{\frac{1}{4N}}^{\frac{3}{4N}}|f'(x)|\int_{N}^{2N}\left(2\pi y-\frac{\pi}{2x}\right)\int_{\frac{1}{2N}}^{x+\frac{1}{4N}}du\,dy\,dx\\
&\quad\quad+\frac{1}{N}\int_{\frac{3}{4N}}^{\infty}|f'(x)|\int_{N}^{2N}\left(2\pi y-\frac{\pi}{2x}\right)\int_{\frac{1}{2N}}^{x+\frac{1}{4N}}du\,dy\,dx\\
&\quad\le C\int_0^{\infty}|f'(x)|\,dx.
\end{aligned}$$

If

$$x\notin\left[u-\frac{1}{4N},u+\frac{1}{4N}\right]$$

we check the integrability of the last integral in (3.46) separately. We now just integrate in the inner integral. Taking 1 in place of $|\sin 2\pi(u-x)y|$, we get the term $(u-x)^{-2}$ to deal with. We obtain

$$\begin{aligned}
\frac{1}{N}\int_{\frac{1}{2N}}^{\infty}\left|\int_{\frac{1}{4N}}^{u-\frac{1}{4N}}\frac{|f'(x)|}{(u-x)^2}\,dx\right|du &\le \frac{1}{N}\int_{\frac{1}{4N}}^{\infty}|f'(x)|\int_{x+\frac{1}{4N}}^{\infty}\frac{1}{(u-x)^2}\,du\,dx\\
&=4\int_0^{\infty}|f'(x)|\,dx.
\end{aligned}$$

Similarly,

$$\begin{aligned}
\frac{1}{N}\int_{\frac{1}{2N}}^{\infty}\left|\int_{u+\frac{1}{4N}}^{\infty}\frac{|f'(x)|}{(u-x)^2}dx\right|du &\le \frac{1}{N}\int_{\frac{3}{4N}}^{\infty}|f'(x)|\int_{\frac{1}{2N}}^{x-\frac{1}{4N}}\frac{1}{(u-x)^2}\,du\,dx\\
&=\frac{1}{N}\int_{\frac{3}{2N}}^{\infty}|f'(x)|\left(4N-\frac{1}{x-\frac{1}{2N}}\right)dx\\
&\le 4\int_0^{\infty}|f'(x)|\,dx.
\end{aligned}$$

It remains to handle (one half of)

$$\left(\int_{\frac{1}{4N}}^{u-\frac{1}{4N}}+\int_{u+\frac{1}{4N}}^{\infty}\right)f'(x)\left[\frac{1}{u-x}-\frac{1}{u+x}\right]\sin\frac{\pi u}{2x}\,dx.$$

Estimating $|\sin\frac{\pi u}{2x}|\le 1$, we thus obtain

$$\begin{aligned}
\int_{\frac{1}{2N}}^{\infty}\int_{\frac{1}{4N}}^{\frac{u}{2}}|f'(x)|\left[\frac{1}{u-x}-\frac{1}{u+x}\right]dx\,du &\le \int_{\frac{1}{4N}}^{\infty}|f'(x)|\int_{2x}^{\infty}\left[\frac{1}{u-x}-\frac{1}{u+x}\right]du\,dx\\
&\le \ln 3\int_0^{\infty}|f'(x)|\,dx.
\end{aligned}$$

Since analogously

$$\int_{\frac{1}{2N}}^{\infty}\int_{\frac{3u}{2}}^{\infty}|f'(x)|\left|\frac{1}{u-x}-\frac{1}{u+x}\right|dx\,du \le \int_{\frac{3}{2N}}^{\infty}|f'(x)|\int_{\frac{1}{2N}}^{\frac{2x}{3}}\left[\frac{1}{u+x}-\frac{1}{u-x}\right]du\,dx$$
$$\le \ln 5\int_0^{\infty}|f'(x)|\,dx,$$

we have to estimate

$$\left(\int_{\frac{u}{2}}^{u-\frac{1}{4N}}+\int_{u+\frac{1}{4N}}^{\frac{3u}{2}}\right)f'(x)\left[\frac{1}{u-x}-\frac{1}{u+x}\right]\sin\frac{\pi u}{2x}\,dx.$$

Further, since

$$\int_{\frac{1}{2N}}^{\infty}\int_{\frac{u}{2}}^{\frac{3u}{2}}|f'(x)|\frac{1}{u+x}\,dx\,du \le \int_0^{\infty}|f'(x)|\int_{2x/3}^{2x}\frac{1}{u-x}\,du\,dx$$
$$= \ln 3\int_0^{\infty}|f'(x)|\,dx$$

and

$$\int_0^{\infty}\int_{\frac{u}{2}}^{\frac{3u}{2}}|f'(x)|\frac{1}{|u-x|}\left|\sin\frac{\pi u}{2x}-1\right|dx\,du \le C\int_0^{\infty}|f'(x)|\int_{\frac{2x}{3}}^{2x}du\,\frac{dx}{x}$$
$$\le C\int_0^{\infty}|f'(x)|\,dx,$$

it remains to estimate

$$\frac{1}{4\pi}\left(\int_{\frac{u}{2}}^{u-\frac{1}{4N}}+\int_{u+\frac{1}{4N}}^{\frac{3u}{2}}\right)\frac{f'(x)}{u-x}\,dx$$

for N large enough. For convenience, we rewrite the last quantity as

$$\frac{1}{4\pi}\left(\int_{\frac{u}{2}}^{u-\varepsilon}+\int_{u+\varepsilon}^{\frac{3u}{2}}\right)\frac{f'(x)}{u-x}\,dx$$

with small ε, or, equivalently,

$$\frac{1}{4\pi}\int_{\substack{\frac{u}{2}\le x\le\frac{3u}{2},\\ |x-u|\ge\varepsilon}}\frac{f'(x)}{u-x}\,dx.$$

By Lemma 3.40, we may consider

$$\frac{1}{4\pi}\int_{|x-u|\ge\varepsilon}\frac{\varphi(x)}{u-x}\,dx$$

rather than the last integral. From now on we no longer need to remember that φ is odd, the only thing we are interested in is the integrability of φ. Unfortunately, considering

$$\int_{\mathbb{R}}\left|\int_{|x-u|\geq\varepsilon}\frac{\varphi(x)}{u-x}\,dx\right|du$$

does not allow one to let ε tend to zero in a proper way and thus does not lead us directly to the desired Hilbert transform. To this end, we wish to consider the Hilbert transform in the discussed above equivalent form. For this, we estimate, for arbitrary $\varepsilon > 0$,

$$\int_{\mathbb{R}}\left|\int_{|x-u|\geq\varepsilon}\frac{\varphi(x)}{u-x}\,dx-\int_{\mathbb{R}}\varphi(x)\frac{u-x}{(u-x)^2+\varepsilon^2}\,dx\right|du.$$

Let us first handle

$$\int_0^\infty\int_{|x-u|\leq\varepsilon}|\varphi(x)|\frac{|u-x|}{(u-x)^2+\varepsilon^2}\,dx\,du$$

(the integral over $(-\infty,0)$ is worked out in exactly the same way). We consider three different cases. The first one is extremely simple:

$$\begin{aligned}\int_0^\varepsilon\int_{|u-x|\leq\varepsilon}|\varphi(x)|\frac{|u-x|}{(u-x)^2+\varepsilon^2}\,dx\,du &\leq \int_0^\varepsilon\int_{-\varepsilon}^\varepsilon|\varphi(x)|\varepsilon^{-1}\,dx\,du\\ &\leq 2\int_{\mathbb{R}}|\varphi(x)|\,dx.\end{aligned}$$

For the second one,

$$\begin{aligned}\int_\varepsilon^{2\varepsilon}\int_{|u-x|\leq\varepsilon}|\varphi(x)|\frac{|u-x|}{(u-x)^2+\varepsilon^2}\,dx\,du &\leq \int_\varepsilon^{2\varepsilon}\int_{u-\varepsilon}^{\varepsilon}|\varphi(x)|\frac{|u-x|}{(u-x)^2+\varepsilon^2}\,dx\,du\\ &\leq \int_0^\varepsilon|\varphi(x)|\int_\varepsilon^{x+\varepsilon}\varepsilon^{-1}du\,dx\leq\int_{\mathbb{R}}|\varphi(x)|\,dx.\end{aligned}$$

Finally,

$$\begin{aligned}\int_\varepsilon^\infty\int_{|u-x|\leq\varepsilon}|\varphi(x)|\frac{|u-x|}{(u-x)^2+\varepsilon^2}\,dx\,dx &= \int_\varepsilon^\infty|\varphi(x)|\int_{x-\varepsilon}^{x+\varepsilon}\frac{|u-x|}{(u-x)^2+\varepsilon^2}\,du\,dx\\ &\leq 2\int_0^\infty|\varphi(x)|\,dx.\end{aligned}$$

We now estimate

$$\int_{\mathbb{R}}\left|\int_{|x-u|\geq\varepsilon}\varphi(x)\left[\frac{1}{u-x}-\frac{u-x}{(u-x)^2+\varepsilon^2}\right]dx\right|du.$$

Since

$$\frac{1}{u-x} - \frac{u-x}{(u-x)^2+\varepsilon^2} = \frac{\varepsilon^2}{(u-x)[(u-x)^2+\varepsilon^2]},$$

dealing again, as above, with u and x both positive, we obtain

$$\begin{aligned}\int_\varepsilon^\infty &\left| \int_0^{u-\varepsilon} \varphi(x) \frac{\varepsilon^2}{(u-x)[(u-x)^2+\varepsilon^2]}\, dx \right| du \\ &\le \int_0^\infty |\varphi(x)|\, \varepsilon \int_{x+\varepsilon}^\infty \frac{du}{(u-x)^2+\varepsilon^2}\, dx \le \frac{\pi}{4} \int_0^\infty |\varphi(x)|\, dx.\end{aligned}$$

It remains to observe that

$$\begin{aligned}\int_0^\infty &\left| \int_{u+\varepsilon}^\infty \varphi(x) \frac{\varepsilon^2}{(u-x)[(u-x)^2+\varepsilon^2]}\, dx \right| du \\ &\le \int_\varepsilon^\infty |\varphi(x)|\, \varepsilon \int_0^{x-\varepsilon} \frac{du}{(u-x)^2+\varepsilon^2}\, dx \le \frac{\pi}{4} \int_0^\infty |\varphi(x)|\, dx.\end{aligned}$$

To complete the proof, we need the following

Lemma 3.47. *Let g be an integrable function. Then*

$$\int_{\mathbb{R}} g(x) \frac{u-x}{(u-x)^2+\varepsilon^2}\, dx = \varepsilon \int_{\mathbb{R}} \frac{\tilde g(x)}{(u-x)^2+\varepsilon^2}\, dx.$$

Proof. This result is proved in [181, Ch. VI, Lemma 1.5] for functions from L^p, $p > 1$, by passing to the Fourier transforms. This idea works here as well but we give a simple direct proof instead. Rewriting the right-hand side and using Fubini's theorem, we have

$$\int_{\mathbb{R}} \frac{1}{(u-x)^2+\varepsilon^2} \frac{1}{\pi} \int_{\mathbb{R}} \frac{g(t)}{x-t}\, dt\, dx = \int_{\mathbb{R}} g(t) \frac{1}{\pi} \int_{\mathbb{R}} \frac{1}{x-t} \frac{1}{(u-x)^2+\varepsilon^2}\, dx\, dt.$$

Substituting $x - u = z\varepsilon$, we obtain

$$\int_{\mathbb{R}} g(t) \frac{1}{\pi} \int_{\mathbb{R}} \frac{1}{x-t} \frac{1}{(u-x)^2+\varepsilon^2}\, dx\, dt = \frac{1}{\varepsilon^2} \int_{\mathbb{R}} g(t) \frac{1}{\pi} \int_{\mathbb{R}} \frac{1}{z - \frac{t-u}{\varepsilon}} \frac{1}{z^2+1}\, dz\, dt.$$

Since

$$\int_{\mathbb{R}} \frac{1}{z-a} \frac{1}{z^2+1}\, dz = \frac{1}{a^2+1} \int_{\mathbb{R}} \left[\frac{1}{z-a} - \frac{z+a}{z^2+1} \right] dz = -\frac{a\pi}{a^2+1},$$

we have $\left(a = \frac{t-u}{\varepsilon}\right)$

$$\begin{aligned}\int_{\mathbb{R}} \frac{\tilde\varphi(x)}{(u-x)^2+\varepsilon^2}\, dx &= \frac{1}{\varepsilon^2} \int_{\mathbb{R}} g(t) \frac{\frac{u-t}{\varepsilon}}{(\frac{u-t}{\varepsilon})^2+1}\, dt \\ &= \frac{1}{\varepsilon} \int_{\mathbb{R}} g(t) \frac{u-t}{(u-t)^2+\varepsilon^2}\, dt,\end{aligned}$$

which completes the proof. □

To complete the proof of the theorem, it remains to apply the lemma with $g = \varphi$ and observe that

$$\int_{\mathbb{R}} \left| \varepsilon \int_{\mathbb{R}} \frac{\tilde{\varphi}(x)}{(u-x)^2+\varepsilon^2}\, dx \right| du \le \int_{\mathbb{R}} |\tilde{\varphi}(x)| \int_{\mathbb{R}} \frac{1}{1+\frac{(u-x)^2}{\varepsilon^2}}\, d\left(\frac{u-x}{\varepsilon}\right) dx$$
$$= \pi \int_{\mathbb{R}} |\tilde{\varphi}(x)|\, dx.$$

The terms estimated above are bounded by $\int_{\mathbb{R}} |\varphi(x)|\, dx$, along with the last one, bounded by $\int_{\mathbb{R}} |\tilde{\varphi}(x)|\, dx$, they can be treated as γ_2. □

3.5.2 The Fourier transform of a function with shifted singularity

The Fourier transform result inspired by Theorem 3.24 is as follows. Let

$$S_f^z = \int_0^\infty \left| \int_0^{\frac{u}{2}} \frac{f'(z+u-x) - f'(z+u+x)}{x} dx \right| du.$$

We obviously have

$$S_f^0 = S_f = \int_0^\infty \left| \int_0^{\frac{u}{2}} \frac{f'(u-x) - f'(u+x)}{x} dx \right| du.$$

For the space of functions of bounded variation, we denote

$$\|f\|_{BV} = \int_0^\infty |f'(x)| dx < \infty.$$

Let also $Q(u,z) = \min(\frac{u}{2}, \frac{z-u}{2})$.

Theorem 3.48. *Let f, g be locally absolutely continuous functions on $(0,\infty)$ vanishing at infinity*

$$\lim_{x\to\infty} f(x) = 0, \qquad \lim_{x\to\infty} g(x) = 0.$$

Then for any $z \ge 0$, $y > 0$,

$$\hat{f}_c(y) + \hat{g}_s(y) = \begin{cases} \frac{1}{2\pi y} g\left(\frac{1}{4y}\right) + \frac{1}{2\pi y} \sin zy \left[f\left(z - \frac{1}{4y}\right) - f\left(z + \frac{1}{4y}\right)\right] \\ \qquad -\frac{1}{2\pi y} \cos zy \left[g\left(z - \frac{1}{4y}\right) - g\left(z + \frac{1}{4y}\right)\right], & y \ge \frac{1}{4z} \\ \qquad \frac{1}{2\pi y} g\left(\frac{1}{4y}\right), & 0 \le y < \frac{1}{4z} \end{cases}$$
$$+\Gamma(y),$$

where

$$\int_0^\infty |\Gamma(y)|dy \le C\Big\{ \|f\|_{BV} + \|g\|_{BV} + S_f^z + S_g^z \\ + \int_0^\infty \left| \int_0^{Q(u,z)} \frac{f'(u-x) - f'(u+x)}{x} dx \right| du \\ + \int_0^\infty \left| \int_0^{Q(u,z)} \frac{g'(u-x) - g'(u+x)}{x} dx \right| du \Big\}.$$

Proof. We first need the following lemma.

Lemma 3.49. *Let g be a locally absolutely continuous function on $(0, \infty)$. Then the following inequality holds:*

$$\int_0^\infty \left| \int_0^{\frac{u}{2}} \frac{g(u-x) - g(u+x)}{x} dx \right| du \le \ln 3 \int_0^\infty t|g'(t)|\, dt.$$

Proof. We have

$$\begin{aligned} \int_0^\infty \left| \int_0^{\frac{u}{2}} \frac{g(u-x) - g(u+x)}{x} dx \right| du &= \int_0^\infty \left| \int_0^{\frac{u}{2}} \frac{dx}{x} \int_{u-x}^{u+x} g'(t)dt \right| du \\ &\le \int_0^\infty \int_{\frac{u}{2}}^{\frac{3u}{2}} |g'(t)| \ln \frac{u}{2|u-t|}\, dt\, du \\ &= \ln 3 \int_0^\infty t|g'(t)|\, dt. \end{aligned}$$

This completes the proof. □

To go further in proving Theorem 3.48, the following generalization (see [78, Lemma 2]) of Lemma 2 in [189] is needed.

Consider two auxiliary functions

$$\beta_f(x) = \begin{cases} f(x), & 0 \le x < \frac{z}{3}, \\ \left(2 - \frac{3x}{z}\right) f(x), & \frac{z}{3} \le x \le \frac{2z}{3}, \\ 0, & x > \frac{2z}{3}, \end{cases}$$

and

$$\gamma_f(x) = \begin{cases} f(z-x) - \beta_f(z-x), & 0 \le x \le z, \\ 0, & x > z. \end{cases}$$

Evidently, $f(x) = \beta_f(x) + \gamma_f(z-x)$ on $[0, z]$.

Lemma 3.50. *Let f be an absolutely continuous function on $[0,z]$. Then the following inequalities hold:*

$$\int_0^\infty (|\beta_f{}'(x)| + |\gamma_f{}'(x)|)\,dx \le C\left(\int_0^z |f'(x)|\,dx + |f(z)|\right) \tag{3.51}$$

and

$$\int_0^\infty \left|\int_0^{\frac{u}{2}} \frac{\beta_f{}'(u-x) - \beta_f{}'(u+x)}{x}dx\right| du + \int_0^\infty \left|\int_0^{\frac{u}{2}} \frac{\gamma_f{}'(u-x) - \gamma_f{}'(u+x)}{x}dx\right| du$$
$$\le C\left(\int_0^z \left|\int_0^{Q(u,z)} \frac{f'(u-x) - f'(u+x)}{x}dx\right| du + \int_0^z |f'(x)|\,dx + |f(z)|\right). \tag{3.52}$$

Proof. Let us write

$$F(x) = \begin{cases} \frac{3}{z}f(x), & \frac{z}{3} \le x \le \frac{2}{3}z, \\ 0, & \text{otherwise.} \end{cases}$$

We are not able to apply Lemma 3.49 to F immediately, because F may be not absolutely continuous in the neighborhoods of $\frac{z}{3}$ and $\frac{2}{3}z$. Let us consider a continuous function $F_\varepsilon(x)$ on $[0,\infty)$ which coincides with F on $\left[\frac{z}{3}, \frac{2}{3}z\right]$, vanishes outside $\left[\frac{z}{3} - \varepsilon, \frac{2}{3}z + \varepsilon\right]$ for sufficiently small ε, and is linear on $\left[\frac{z}{3} - \varepsilon, \frac{z}{3}\right]$ and $\left[\frac{2}{3}z, \frac{2}{3}z + \varepsilon\right]$. Since F_ε satisfies the conditions of Lemma 3.49, we obtain

$$\begin{aligned}
&\int_0^\infty \left|\int_0^{\frac{u}{2}} \frac{F_\varepsilon(u-x) - F_\varepsilon(u+x)}{x}dx\right| du \\
&\le \ln 3 \int_0^\infty t|F_\varepsilon'(t)|\,dt \le \ln 3 \left\{\frac{3}{z}\int_{\frac{z}{3}}^{\frac{2}{3}z} t|f'(t)|\,dt + \left|f\left(\frac{z}{3}\right)\right| + \left|f\left(\frac{2}{3}z\right)\right|\right\} \\
&\le \ln 3\left(2\int_{\frac{z}{3}}^{\frac{2}{3}z} |f'(t)|\,dt + \left|\int_{\frac{z}{3}}^z f'(t)dt - f(z)\right| + \left|\int_{\frac{2z}{3}}^z f'(t)dt - f(z)\right|\right) \\
&\le 3\ln 3\left(\int_0^z |f'(t)|\,dt + |f(z)|\right).
\end{aligned} \tag{3.53}$$

Further, one can easily calculate that

$$\begin{aligned}
&\int_0^\infty \left|\int_0^{\frac{u}{2}} \frac{(F - F_\varepsilon)(u-x) - (F - F_\varepsilon)(u+x)}{x}dx\right| du \\
&\le C\left(\left|f\left(\frac{z}{3}\right)\right| + \left|f\left(\frac{2}{3}z\right)\right|\right) \le C\left(\int_0^z |f'(t)|dt + |f(z)|\right).
\end{aligned} \tag{3.54}$$

Thus, we derive from (3.53) and (3.54) that

$$\int_0^\infty \left| \int_0^{\frac{u}{2}} \frac{F(u-x)-F(u+x)}{x}\, dx \right| du \le C \left(\int_0^z |f'(t)|\, dt + |f(z)| \right). \tag{3.55}$$

Let us write $B'(x) = \beta'(x) + F(x)$, *i.e.*,

$$B'(x) = \begin{cases} f'(x), & 0 \le x < \frac{z}{3}, \\ \left(2 - \frac{3x}{z}\right) f'(x), & \frac{z}{3} \le x \le \frac{2z}{3}, \\ 0, & x > \frac{2z}{3}. \end{cases} \tag{3.56}$$

We get from (3.55) and (3.56)

$$\begin{aligned} &\int_0^\infty \left| \int_0^{\frac{u}{2}} \frac{\beta'(u-x)-\beta'(u+x)}{x}\, dx \right| du \\ &\le \int_0^\infty \left| \int_0^{\frac{u}{2}} \frac{B'(u-x)-B'(u+x)}{x} dx \right| du + C \left(\int_0^z |f'(x)|\, dx + |f(z)| \right). \end{aligned} \tag{3.57}$$

It follows from (3.56) that $B'(u-x) = f'(u-x)$ for $u \le \frac{z}{3}$, and

$$B'(u+x) = \begin{cases} f'(u+x), & x \le \frac{z}{3} - u, \\ \left(2 - \frac{3(u+x)}{z}\right) f'(u+x), & x > \frac{z}{3} - u. \end{cases}$$

Therefore,

$$\begin{aligned} &\int_0^{\frac{z}{3}} \left| \int_0^{\frac{u}{2}} \frac{(B'-f')(u-x)-(B'-f')(u+x)}{x} dx \right| du \\ &= \int_0^{\frac{z}{3}} \left| \int_{\frac{z}{3}-u}^{\frac{u}{2}} \left(\frac{3}{z}(u+x) - 1 \right) \frac{f'(u+x)}{x}\, dx \right| du \le \int_0^z |f'(x)|\, dx. \end{aligned} \tag{3.58}$$

Let $\frac{z}{3} \le u \le \frac{2}{3}z$. Then it follows from (3.56) that

$$B'(u-x) = \begin{cases} \left(2 - \frac{3(u-x)}{z}\right) f'(u-x), & x \le u - \frac{z}{3}, \\ f'(u-x), & x > u - \frac{z}{3}, \end{cases}$$

and

$$B'(u+x) = \begin{cases} \left(2 - \frac{3(u+x)}{z}\right) f'(u+x), & x < \frac{2}{3}z - u, \\ 0, & x \ge \frac{2}{3}z - u. \end{cases}$$

This yields

$$
\begin{aligned}
&\int_{\frac{z}{3}}^{\frac{2z}{3}}\left|\int_0^{\frac{u}{2}} \frac{B'(u-x)-B'(u+x)}{x}\,dx - \left(2-\frac{3u}{z}\right)\int_0^{\frac{u}{2}} \frac{f'(u-x)-f'(u+x)}{x}\,dx\right| du \\
&= \int_{\frac{z}{3}}^{\frac{2z}{3}} \left| \int_0^{u-\frac{z}{2}} \frac{3f'(u-x)}{z}\,dx \right. \\
&\quad + \left(\frac{3u}{z}-1\right)\int_{u-\frac{z}{3}}^{\frac{u}{2}} \frac{f'(u-x)}{x}\,dx + \int_0^{\min\left(\frac{u}{2},\frac{2z}{3}-u\right)} \frac{3f'(u+x)}{z}\,dx \\
&\quad \left. + \left(2-\frac{3u}{z}\right)\int_{\frac{2z}{3}-u}^{\frac{u}{2}} \frac{f'(u+x)}{x}\,dx \right| du \\
&\le \frac{3}{z}\int_{\frac{z}{3}}^{\frac{2z}{3}} \left\{ \int_0^{\frac{u}{2}} |f'(u-x)|\,dx + \int_0^{\frac{u}{2}} |f'(u+x)|\,dx \right\} du \le \int_0^z |f'(x)|\,dx.
\end{aligned}
$$

Thus, we have obtained

$$
\begin{aligned}
&\int_{\frac{z}{3}}^{\frac{2z}{3}} \left| \int_0^{\frac{u}{2}} \frac{B'(u-x)-B'(u+x)}{x}\,dx \right| du \\
&\quad \le \int_{\frac{z}{3}}^{\frac{2z}{3}} \left| \int_0^{\frac{u}{2}} \frac{f'(u-x)-f'(u+x)}{x}\,dx \right| du + \int_0^z |f'(x)|\,dx.
\end{aligned}
\tag{3.59}
$$

Let $u \ge \frac{2z}{3}$. The formula (3.56) gives us that $B'(u+x)=0$ and

$$
B'(u-x) = \begin{cases} 0, & x \le u - \frac{2}{3}z \\ \left(2-\frac{3(u-x)}{z}\right) f'(u-x), & x > u - \frac{2}{3}z. \end{cases}
$$

Hence

$$
\begin{aligned}
&\int_{\frac{2}{3}z}^{\infty} \left| \int_0^{\frac{u}{2}} \frac{B'(u-x)-B'(u+x)}{x}\,dx \right| du \\
&\quad = \int_{\frac{2}{3}z}^{\infty} \left| \int_{u-\frac{2}{3}z}^{\frac{u}{2}} \left(2-\frac{3(u-x)}{z}\right) \frac{f'(u-x)}{x} \right| du \\
&\quad \le \frac{3}{z}\int_{\frac{2}{3}z}^{\frac{4z}{3}} du \int_{u-\frac{2}{3}z}^{\frac{u}{2}} |f'(u-x)|\,dx \le 2\int_0^z |f'(x)|\,dx.
\end{aligned}
\tag{3.60}
$$

Collecting the estimates (3.58)–(3.60), we have

$$
\begin{aligned}
&\int_0^{\infty} \left| \int_0^{\frac{u}{2}} \frac{B'(u-x)-B'(u+x)}{x}\,dx \right| du \\
&\quad \le \int_0^{\frac{2z}{3}} \left| \int_0^{\frac{u}{2}} \frac{f'(u-x)-f'(u+x)}{x}\,dx \right| du + 4\int_0^z |f'(x)|\,dx.
\end{aligned}
$$

If $u > \frac{z}{2}$, then

$$\int_{\frac{z}{2}}^{\frac{2z}{3}} \left| \int_{\frac{z-u}{2}}^{\frac{u}{2}} \frac{f'(u-x) - f'(u+x)}{x} dx \right| du$$
$$\leq \frac{3}{z} \int_{\frac{z}{2}}^{\frac{2z}{3}} du \int_{\frac{z-u}{2}}^{\frac{u}{2}} (|f'(u-x)| + |f'(u+x)| dx \leq \int_0^z |f'(x)|\, dx.$$

So we have

$$\int_0^\infty \left| \int_0^{\frac{u}{2}} \frac{B'(u-x) - B'(u+x)}{x} dx \right| du$$
$$\leq \int_0^{\frac{2}{3}z} \left| \int_0^{\min\left(\frac{u}{2}, \frac{z-u}{2}\right)} \frac{f'(u-x) - f'(u+x)}{x}\, dx \right| du + 5 \int_0^z |f'(x)|\, dx.$$

Taking into account (3.57), we see that the inequality

$$\int_0^\infty \left| \int_0^{\frac{u}{2}} \frac{\beta'(u-x) - \beta'(u+x)}{x} dx \right| du$$
$$\leq \int_0^{\frac{2z}{3}} \left| \int_0^{\min\left(\frac{u}{2}, \frac{z-u}{2}\right)} \frac{f'(u-x) - f'(u+x)}{x}\, dx \right| du + C \left(\int_0^z |f'(x)|\, dx + |f(z)| \right)$$

holds. We now have

$$\int_0^\infty |\beta'(x)|\, dx \leq \int_0^{\frac{z}{3}} |f'(x)|\, dx + \int_{\frac{z}{3}}^{\frac{2z}{3}} \left(2 - \frac{3x}{z}\right) |f(x)|\, dx + \frac{3}{z} \int_{\frac{z}{3}}^{\frac{2z}{3}} |f(x)|\, dx$$
$$\leq 2 \int_0^z |f'(x)|\, dx + |f(x)|.$$

We have now proved (3.51) and (3.52) for β. The corresponding estimates for γ are similar to those for β. Lemma 3.50 is proved. □

With this result in hand we are ready to continue proving the theorem. For $\widehat{f_c}(y)$, we use (3.38) for $0 < y < \frac{1}{4z}$, while for $y \geq \frac{1}{4z}$ we use Theorem 3.37. Let us go on to $\widehat{g}_s(y)$. For $0 < y < \frac{1}{4z}$ nothing remains but to use (3.39). We have for $y \geq \frac{\pi}{2z}$,

$$\int_0^\infty g(x) \sin 2\pi xy\, dx = \int_0^z g(x) \sin 2\pi xy\, dx + \int_z^\infty g(x) \sin 2\pi xy\, dx.$$

For the last integral, we get

$$\int_z^\infty g(x) \sin 2\pi xy\, dx = \int_0^\infty g(z+x) \sin 2\pi(z+x)y\, dx$$
$$= \sin 2\pi zy \int_0^\infty g(z+x) \cos 2\pi xy\, dx + \cos 2\pi zy \int_0^\infty g(z+x) \sin 2\pi xy\, dx,$$

and it suffices to apply Theorem 3.37 to both integrals. All is clear with the first one, while applying (3.39) to the last integral yields

$$\cos 2\pi zy \int_0^\infty g(z+x)\sin 2\pi xy\,dx = \frac{1}{2\pi y} g\left(z+\frac{1}{4y}\right)\cos 2\pi zy + \Gamma(y).$$

Further,

$$\int_0^z g(x)\sin 2\pi xy\,dx = \int_0^z \beta_g(x)\sin 2\pi xy\,dx + \int_0^z \gamma_g(z-x)\sin 2\pi xy\,dx.$$

Now, we apply (3.39) to the first integral on the right-hand side and obtain

$$\int_0^z \beta_g(x)\sin 2\pi xy\,dx = \frac{1}{2\pi y}\beta_g\left(\frac{1}{4y}\right) + \Gamma(y). \tag{3.61}$$

Similarly, for the second one, we have

$$\begin{aligned}\int_0^z \gamma_g(z-x)\sin 2\pi xy\,dx &= \int_0^z \gamma_g(x)\sin 2\pi(z-x)y\,dx\\ &= \sin 2\pi zy \int_0^z \gamma_g(x)\cos 2\pi xy\,dx - \cos 2\pi zy\int_0^z \gamma_g(x)\sin 2\pi xy\,dx\\ &= -\frac{1}{2\pi y}\cos 2\pi zy\,\gamma_g\left(\frac{1}{4y}\right) + \Gamma(y).\end{aligned}$$

Since $\frac{1}{4y} \le z$, we have

$$\begin{aligned}&\int_0^z \gamma_g(z-x)\sin 2\pi xy\,dx\\ &= -\frac{1}{2\pi y}\cos 2\pi zy\, g\left(z-\frac{1}{4y}\right) + \frac{1}{2\pi y}\cos 2\pi zy\,\beta_g\left(z-\frac{1}{4y}\right) + \Gamma(y).\end{aligned} \tag{3.62}$$

Recalling the definition of β_g and combining (3.61) and (3.62) yields that $\widehat{g}_s(y)$ is

$$\frac{1}{2\pi y}\cos 2\pi zy\, g\left(z+\frac{1}{4y}\right)$$

for $\frac{1}{4z} \le y < \frac{3}{8z}$,

$$\begin{aligned}&\frac{1}{\pi y} g\left(\frac{1}{4y}\right) - \frac{1}{2\pi y}\cos 2\pi zy\, g\left(z+\frac{1}{4y}\right)\\ &\quad + \frac{1}{\pi y}\cos 2\pi zy\left[g\left(z+\frac{1}{4y}\right) - g\left(z-\frac{1}{4y}\right)\right]\end{aligned}$$

for $\frac{3}{8z} \le y < \frac{3}{4z}$, and

$$\frac{1}{2\pi y} g\left(\frac{1}{4y}\right) - \frac{1}{2\pi y}\cos 2\pi zy\left[g\left(z-\frac{1}{4y}\right) - g\left(z+\frac{1}{4y}\right)\right]$$

for $y \ge \frac{3}{4z}$, plus the remainder term $\Gamma(y)$. Observe that each term for

$$\frac{1}{4z} \le y < \frac{3}{8z} \quad \text{and} \quad \frac{3}{8z} \le y < \frac{3}{4z},$$

respectively, is easily estimated by $\|g\|_{BV}$ or $\|f\|_{BV}$. Indeed, for example, we have

$$\begin{aligned}\int_{\frac{p}{4z}}^{\frac{q}{4z}} \frac{1}{y}\left|g\left(z-\frac{1}{4y}\right)\right| dy &\le \int_{\frac{p}{4z}}^{\frac{q}{4z}} \left|\int_{\frac{z}{q}}^{z-\frac{1}{4y}} g'(t)\,dt\right| \frac{dy}{y} + \int_{\frac{p}{4z}}^{\frac{q}{4z}} \left|\int_{\frac{z}{q}}^{\infty} g'(t)\,dt\right| \frac{dy}{y} \\ &\le \int_{\frac{p}{4z}}^{\frac{q}{4z}} \int_{\frac{z}{q}}^{z-\frac{1}{4y}} |g'(t)|\,dt\,\frac{dy}{y} + \ln\left(\frac{q}{p}\right)\|g\|_{BV} \le C\|g\|_{BV}.\end{aligned}$$

We already have $\frac{1}{y}g\left(\frac{1}{4y}\right)$ as the leading term of the asymptotic formula in Theorem 3.48 for all y except

$$\frac{1}{4z} \le y < \frac{3}{8z}.$$

It remains to observe that

$$\int_{\frac{1}{4z}}^{\frac{3}{8z}} \left|g\left(\frac{1}{4y}\right)\right| \frac{dy}{y} \le C\|g\|_{BV}$$

as above. To complete the proof, it remains to consider the part in the leading term over

$$\frac{1}{4z} \le y < \frac{3}{4z}.$$

Since in the above argument p and q are arbitrary, it is controlled either by $\|g\|_{BV}$ or by $\|f\|_{BV}$. The proof is complete. □

3.5.3 Amalgam type spaces

Let us now prove similar results for the amalgam type space $A_{1,2}$ in a slightly different form, see [128]. We study, for $\gamma = 0$ or 1, the Fourier transforms

$$\widehat{f}_\gamma(x) = \int_0^\infty f(t)\cos 2\pi\left(xt - \frac{\gamma}{4}\right) dt. \tag{3.63}$$

It is clear that $\widehat{f}_\gamma$ represents the cosine Fourier transform in the case $\gamma = 0$, while taking $\gamma = 1$ gives the sine Fourier transform.

Theorem 3.64. *Let f be locally absolutely continuous on $(0, \infty)$ and vanishing at infinity, that is, $\lim_{t\to\infty} f(t) = 0$, and $f' \in A_{1,2}$. Then for $x > 0$,*

$$\widehat{f}_\gamma(x) = \frac{1}{2\pi x} f\left(\frac{1}{4x}\right) \sin\frac{\pi\gamma}{2} + \Gamma(x),$$

where $\gamma = 0$ or 1, and $\|\Gamma\|_{L^1(\mathbb{R}_+)} \lesssim \|f'\|_{A_{1,2}}$.

We will see that in order to control the L^1 norm of the Fourier transform, no matter cosine or sine, of a function of bounded variation by means of the $A_{1,2}$ norm, the crucial role belongs to the bounds of a special sequence of integrals over the dyadic intervals $[2^m, 2^{m+1}]$. Given an integrable function g, we define the sequence of functions

$$\widehat{G_m}(x) = \int_{2^{-m}}^{\infty} g(t)e^{-2\pi ixt}\,dt.$$

Obviously, this function is the Fourier transform of the function $G_m(t)$ which is $g(t)$ for $2^{-m} < t < \infty$ and zero otherwise.

The above-mentioned integrals are estimated in the next lemma the statement and the proof of which are inspired by Lemma 2 in [6].

Lemma 3.65. *Let g be an integrable function on $\mathbb{R}_+$. Then for $m = 0, \pm1, \pm2, \dots$*

$$\int_{2^m}^{2^{m+1}} \frac{|\widehat{G_m}(x)|}{x}\,dx \ll \left(\sum_{j=1}^{\infty}\left[\int_{j2^{-m}}^{(j+1)2^{-m}} |g(t)|\,dt\right]^2\right)^{\frac12}.$$

Proof of Lemma 3.65. We start with the following inequality:

$$\int_{2^m}^{2^{m+1}} \frac{|\widehat{G_m}(x)|}{x}\,dx \lesssim \int_{2^m}^{2^{m+1}} |\widehat{S_{2^{-m}}}(x)\,\widehat{G_m}(x)|\,dx, \tag{3.66}$$

where

$$\widehat{S_a}(x) = \frac{\sin ax}{x}, \qquad a > 0.$$

The latter can be considered as the Fourier transform, up to a constant, of the indicator function of the interval $[0, a]$. It follows from the formula (see (5) in [31, Ch. I, §4]; it is mentioned in Remark 12 in the cited literature of [31] that the formula goes back to Fourier)

$$\int_0^{\infty} \frac{\sin ax}{x}\cos yx\,dx = \begin{cases} \frac{\pi}{2}, & y < a; \\ \frac{\pi}{4}, & y = a; \\ 0, & y > a. \end{cases}$$

By the Bunyakovskii–Schwarz–Cauchy inequality, the right-hand side of (3.66) does not exceed

$$C2^{\frac{m}{2}}\left(\int_{2^m}^{2^{m+1}} |\widehat{S_{2^{-m}}}(x)\widehat{G_m}(x)|^2\,dx\right)^{\frac12}.$$

In fact, we no longer need the integral over $[2^m, 2^{m+1}]$ (we got the factor $2^{\frac{m}{2}}$ from it) and have to estimate

$$2^{\frac{m}{2}}\left(\int_{\mathbb{R}} |\widehat{S_{2^{-m}}}(x)\,\widehat{G_m}(x)|^2\,dx\right)^{\frac12}.$$

Since G_m is integrable and $S_{2^{-m}}$ is square integrable, $\widehat{S_{2^{-m}}}(x)\,\widehat{G_m}(x)$ is the Fourier transform of their convolution, and both are square integrable; see Theorems 64 and 65 in [193, 3.13]. By Parseval's identity, we estimate

$$2^{\frac{m}{2}}\left(\int_{\mathbb{R}} |(S_{2^{-m}} * G_m)(x)|^2\,dx\right)^{\frac{1}{2}}. \tag{3.67}$$

Further,

$$\widehat{G_m}(x) = \sum_{j=1}^{\infty} \int_{j2^{-m}}^{(j+1)2^{-m}} g(t)e^{-ixt}\,dt = \sum_{j=1}^{\infty} \widehat{G_{m,j}}(x),$$

where

$$\widehat{G_{m,j}}(x) = \int_{j2^{-m}}^{(j+1)2^{-m}} g(t)e^{-ixt}\,dt.$$

Correspondingly,

$$G_m(x) = \sum_{j=1}^{\infty} g_{m,j}(x),$$

with $g_{m,j}(x) = g(x)$ when

$$j2^{-m} \le x < (j+1)2^{-m}$$

and zero otherwise. Representing (3.67) as

$$2^{\frac{m}{2}}\left(\int_{\mathbb{R}} \left|\sum_{j=1}^{\infty} S_{2^{-m}} * g_{m,j}(x)\right|^2 dx\right)^{\frac{1}{2}},$$

let us analyze what the support of each summand

$$S_{2^{-m}} * g_{m,j}(x) = \int_{j2^{-m}}^{(j+1)2^{-m}} S_{2^{-m}}(x-t)g(t)\,dt$$

is. Since we have $0 < x - t < 2^{-m}$, such a summand is supported within the interval

$$j2^{-m} \le x \le (j+2)2^{-m}.$$

Only two neighboring intervals may have an intersection of positive measure. Therefore, the value in (3.67) is dominated by

$$2^{\frac{m}{2}}\left(\sum_{\substack{j=1\\ j \text{ is even}}}^{\infty} \int_{\mathbb{R}} |S_{2^{-m}} * g_{m,j}(x)|^2\,dx\right)^{\frac{1}{2}} + 2^{\frac{m}{2}}\left(\sum_{\substack{j=1\\ j \text{ is odd}}}^{\infty} \int_{\mathbb{R}} |S_{2^{-m}} * g_{m,j}(x)|^2\,dx\right)^{\frac{1}{2}}.$$

The bound for each of the two values is the same and can be obtained by means of Young's inequality for convolution (see (1.11)). Taking $\varphi = S_{2^{-m}}$ and $\psi = g_{m,j}$, $q = 1$ and $p = r = 2$, we obtain in each of the two cases

$$2^{\frac{m}{2}}\left(\sum_{j=1}^{\infty} \|S_{2^{-m}}\|_2^2 \, \|g_{m,j}\|_1^2\right)^{\frac{1}{2}}.$$

Since

$$\|S_{2^{-m}}\|_2^2 \lesssim \int_0^{2^{-m}} dx \lesssim 2^{-m},$$

we get the required bound

$$\left(\sum_{j=1}^{\infty}\left[\int_{j2^{-m}}^{(j+1)2^{-m}} |g(t)|\, dt\right]^2\right)^{\frac{1}{2}}.$$

This completes the proof of the lemma. □

Now, the proof of the theorem runs similarly to that of Theorem 3.37.

Proof of Theorem 3.64. Splitting the integral in (3.63) and integrating by parts, we obtain

$$\begin{aligned}\widehat{f}_\gamma(x) = &-\frac{1}{2\pi x} f\left(\frac{1}{4x}\right)\sin\frac{\pi}{2}(1-\gamma) + \int_0^{\frac{1}{4x}} f(t)\cos 2\pi\left(xt - \frac{\gamma}{4}\right) dt \\ &- \frac{1}{2\pi x}\int_{\frac{1}{4x}}^{\infty} f'(t)\sin 2\pi\left(xt-\frac{\gamma}{4}\right) dt.\end{aligned}$$

Further,

$$\begin{aligned}&\int_0^{\frac{1}{4x}} f(t)\cos 2\pi\left(xt-\frac{\gamma}{4}\right) dt \\ &= \int_0^{\frac{1}{4x}}\left[f(t) - f\left(\frac{1}{4x}\right)\right]\cos 2\pi\left(xt-\frac{\gamma}{4}\right) dt \\ &\quad + \int_0^{\frac{1}{4x}} f\left(\frac{1}{4x}\right)\cos 2\pi\left(xt-\frac{\gamma}{4}\right) dt \\ &= -\int_0^{\frac{1}{4x}}\left[\int_t^{\frac{1}{4x}} f'(s)\, ds\right]\cos 2\pi\left(xt-\frac{\gamma}{4}\right) dt \\ &\quad + \frac{1}{2\pi x} f\left(\frac{1}{4x}\right)\sin\frac{\pi}{2}(1-\gamma) + \frac{1}{2\pi x} f\left(\frac{1}{4x}\right)\sin\frac{\pi\gamma}{2} \\ &= \frac{1}{2\pi x} f\left(\frac{1}{4x}\right)\sin\frac{\pi\gamma}{2} + \frac{1}{2\pi x} f\left(\frac{1}{4x}\right)\sin\frac{\pi}{2}(1-\gamma) + O\left(\int_0^{\frac{1}{4x}} s|f'(s)|\, ds\right).\end{aligned}$$

Since

$$\int_0^\infty \int_0^{\frac{1}{4x}} s|f'(s)|\,ds\,dx = \frac{1}{4}\int_0^\infty |f'(s)|\,ds,$$

it follows from (3.33) and the remark thereafter that to prove the theorem it remains to estimate

$$\int_0^\infty \frac{1}{x}\left|\int_{\frac{1}{4x}}^\infty f'(t)\sin 2\pi\left(xt-\frac{\gamma}{4}\right)dt\right| dx.$$

We can study

$$\sum_{m=-\infty}^{\infty}\int_{2^m}^{2^{m+1}} \frac{1}{x}\left|\int_{2^{-m}}^\infty f'(t)\sin 2\pi\left(xt-\frac{\gamma}{4}\right)dt\right| dx$$

instead. Indeed,

$$\sum_{m=-\infty}^{\infty}\int_{2^m}^{2^{m+1}} \frac{1}{x}\left|\int_{2^{-m}}^{\frac{1}{4x}} f'(t)\sin 2\pi\left(xt-\frac{\gamma}{4}\right)dt\right| dx$$

$$\le \sum_{m=-\infty}^{\infty}\int_{2^m}^{2^{m+1}} \frac{1}{x}\int_{\frac{1}{4x}}^{\frac{1}{x}} |f'(t)|\,dt\,dx \lesssim \int_0^\infty |f'(t)|\,dt.$$

Applying now the proven Lemma 3.65, we complete the proof of the theorem, since m runs from $-\infty$ to ∞ and we can write 2^m instead of 2^{-m}. □

3.6 Applications and further progress

Theorem 3.34 and its consequences has numerous applications. First of all it allows us to strengthen known results on the integrability of trigonometric series. Given the series (3.19) or (3.20) with the null sequence of coefficients in χ, set for $x \in [k-1,k]$,

$$A(x) = a_k + (k-x)\Delta a_{k-1}, \qquad a_0 = 0,$$
$$B(x) = b_k + (k-x)\Delta b_{k-1}.$$

The following result due to Trigub [201, Theorem 4] (see also [203]) is a "bridge" between sequences of Fourier coefficients and Fourier transforms (for an extension, see a recent paper [202]; an earlier version, for functions with compact support, is due to Belinsky [19]):

$$\sup_{0<|y|\le\frac{1}{2}}\left|\int_{-\infty}^{+\infty}\varphi(x)e^{2\pi ixy}\,dx - \sum_{-\infty}^{+\infty}\varphi(k)e^{2\pi iky}\right|$$
$$\le C\|\varphi\|_{BV}. \tag{3.68}$$

This is, in a sense, equiconvergence of the Fourier integral and trigonometric series, both generated by a function of bounded variation.

Corollary 3.69. *For each* y, $0 < y \le \frac{1}{2}$,

$$\sum_{k=1}^{\infty} a_k \cos 2\pi k y = \gamma(y),$$

where

$$\int_0^{\pi} |\gamma(y)|\, dy \le C(\|a\|_{bv} + s_a),$$

$$\sum_{k=1}^{\infty} b_k \sin 2\pi k y = \frac{1}{2\pi y} B\left(\frac{1}{4y}\right) + \gamma(y),$$

where

$$\int_0^{\pi} |\gamma(y)|\, dy \le C(\|b\|_{bv} + s_b).$$

This immediately yields Theorem 3.22, merely by integrating the formulas obtained. Observe that besides direct answers to the above question, estimates of the integral

$$I = \int \left| \frac{a_0}{2} + \sum_{k=1}^{\infty} (a_k \cos 2\pi k x + b_k \sin 2\pi k x) \right| dx$$

over some smaller interval may be helpful.

We can now obtain not only Theorem 3.24 but a stronger version, in the spirit of Corollary 3.69.

Theorem 3.70. *Let* $\{a_k\}$ *and* $\{b_k\}$ *be null sequences. Then for any* y, $0 < y \le \frac{1}{2}$ *and* $z \in [m, m+1]$, $m = 0, 1, 2, \ldots$, *we have*

$$\sum_{k=1}^{\infty} (a_k \cos 2\pi k y + b_k \sin 2\pi k y) = \begin{cases} \begin{aligned} &\tfrac{1}{2\pi y} B\left(\tfrac{1}{4y}\right) \\ &+ \tfrac{1}{2\pi y} \sin 2\pi z y \left[A\left(z - \tfrac{1}{4y}\right) \right. \\ &\left. - A\left(z + \tfrac{1}{4y}\right) \right] \\ &- \tfrac{1}{2\pi y} \cos 2\pi z y \left[B\left(z - \tfrac{1}{4y}\right) \right. \\ &\left. - B\left(z + \tfrac{1}{4y}\right) \right], \end{aligned} & y \ge \frac{1}{4z} \\ \tfrac{1}{2\pi y} B\left(\tfrac{1}{4y}\right), & 0 \le y < \frac{1}{4z} \end{cases}$$

$$+ \gamma(y),$$

where

$$\int_0^{\frac{1}{2}} |\gamma(y)|\,dy \le C\left(\|a\|_{bv} + \|b\|_{bv} + s_a^m + s_b^m + \sum_{n=2}^{m-2}\left|\sum_{k=1}^{q_{n,m}} \frac{\Delta a_{n-k} - \Delta a_{n+k}}{k}\right|\right.$$
$$\left. + \sum_{n=2}^{m-2}\left|\sum_{k=1}^{q_{n,m}} \frac{\Delta b_{n-k} - \Delta b_{n+k}}{k}\right|\right).$$

In the same manner as above a similar result for the amalgam type space can be derived, slightly more general than that from [6].

Theorem 3.71. *If the coefficients* $\{a_n\}$ *in* (3.19) *and* $\{b_n\}$ *in* (3.20) *tend to* 0 *as* $n \to \infty$, *and the sequences* $\{\Delta a_n\}$ *and* $\{\Delta b_n\}$ *are in* $a_{1,2}$, *then* (3.19) *represents an integrable function on* $[0, \frac{1}{2}]$, *and*

$$\sum_{n=1}^{\infty} b_n \sin nx = \frac{1}{x} B\left(\frac{\pi}{2x}\right) + \Gamma(x), \tag{3.72}$$

where

$$\int_0^{\frac{1}{2}} |\Gamma(x)|\,dx \lesssim \|\{\Delta b_n\}\|_{a_{1,2}}.$$

Back to the mentioned results by Bausov [13] and Telyakovskii [191], they and their generalizations were applied in the work of Ganzburg and one of the authors to obtain best approximation for infinitely differentiable functions (see, *e.g.*, [78]).

Not going into details why usual methods are not applicable to infinitely differentiable functions, we will just outline the idea how the Fourier transform comes into play.

Taking F to be a continuous function with integrable cosine Fourier transform and satisfying some conditions relevant to the considered stuff, one introduces the operator

$$Q_\sigma(F, y) = 2\int_0^{\sigma} (\hat{F}_c(x) - \hat{F}_c(2\sigma - x)) \cos 2\pi xy\,dx.$$

For each F this is an entire function of type σ. Writing

$$f(x) = \begin{cases} \hat{F}_c(2\sigma - x), & 0 \le x \le \sigma; \\ \hat{F}_c(x), & x > \sigma, \end{cases}$$

we observe that the identity holds

$$F(x) - Q_\sigma(F, x) = 2\hat{f}_c(x).$$

Then the obtained results on the Fourier transform and those related allow one to get sharp estimates for the best approximation by entire functions in integral metrics.

Applications to estimates of Lebesgue constants are known. Note also the results on multipliers in [84, 85] and [123].

Theorem 3.34 (or, equivalently, Theorem 3.37) and its related conclusions appear to be in a final form, even maybe the strongest possible. However, the actual situation is quite different. In recent papers [127] and [129], the widest possible spaces for integrability of the cosine and sine Fourier transforms are given. In particular, this immediately proves the first part of Theorem 3.37. However, the situation with the sine Fourier transform is in any case much more delicate. These seem to open a new page in the study of integrability of the Fourier transform and, consequently, in integrability of trigonometric series.

3.7 Multivariate case

We now come to the multidimensional case. There exist generalizations of the mentioned one-dimensional results. However, it is worth mentioning that the above one-dimensional results can be applied in the multidimensional setting. The point is that because of the Fourier slice theorem (2.28) the multidimensional Fourier transform can be represented via the one-dimensional one of the Radon transform. The problem here is to pose appropriate assumptions on the function and figure out how the Radon transform inherits its properties. A different way for direct application of the one-dimensional results concerns radial functions and will be discussed in detail in the next chapter.

As in dimension one, a good amount of results for the multidimensional Fourier transform come as generalizations of the results on the integrability of trigonometric series. An outline of one-dimensional results on the integrability of trigonometric series is given above. The number of corresponding multidimensional extensions may be compared with the number of the most important one-dimensional results. It is natural, in a sense, since such extensions are mostly proved by repeating the corresponding one-dimensional arguments. Nevertheless, it is not always so simple as it may seem, and sometimes peculiarities of the multidimensional case are displayed. One can find more details in [125]. Here we give only some related results and illustrate how they can be used.

Let

$$I(q,p), \quad 0 \le q \le d, \quad 1 \le p \le \binom{d}{q},$$

be the pth subset from all possible different subsets of I consisting of $d-q$ elements; and let

$$I(q,p;s,r), \quad 0 \le s \le d-q, \quad 1 \le r \le \binom{d-q}{s},$$

be the rth one from all possible different subsets of $I(q,p)$ consisting of $n-q-s$ elements. We denote by $\partial_{q,p} f$ the partial derivative of a function f taken with respect to every variable with index from $I(q,p)$. Given a function φ defined on $\mathbb{R}_+$, let $\varphi_{s,r}$ denote the odd extension of φ in each variable with index in $I(q,p;s,r)$.

Theorem 3.73. *Let f be defined on $\mathbb{R}^d_+$; let for each q,p, $1\le q\le d$, $1\le p\le \binom{d}{q}$, the functions $\partial_{q,p}f$ be locally absolutely continuous with respect to every variable with index from $I\setminus I(q,p)$ and*

$$\lim_{x_1+\cdots+x_d\to\infty}\partial_{q,p}f=0.$$

Then for any $y_1,\dots,y_d>0$ and for any set of numbers $\{a_j : a_j=0 \text{ or } \frac14, j\in I\}$ we have

$$\int_{\mathbb{R}^d_+} f(x)\prod_{j=1}^d \sin 2\pi(x_jy_j+a_j)\,dx_j=(-1)^d f\left(\frac{1}{4y_1},\dots,\frac{1}{4y_d}\right)\prod_{j=1}^d\frac{1-4a_j}{2\pi y_j}+\gamma(y), \tag{3.74}$$

where

$$\int_{\mathbb{R}^d_+}|\gamma(y)|\,dy\le C\sum_{q=0}^{d-1}\sum_{p=1}^{\binom{d}{k}}\sum_{s=0}^{d-q}\sum_{r=1}^{\binom{d-q}{s}}\int_{\mathbb{R}_+}\Bigg|\int_{\mathbb{R}^{d-q-s}}(\partial_{q,p}f)_{s,r}(x^y_{q,p})\times\prod_{j\in I(q,p;s,r)}\frac{dx_j}{y_j-x_j}\Bigg|\prod_{j\in I\setminus I(q,p)}\frac{\cos 2\pi a_j}{y_j}\,dy,$$

provided the right-hand side of the last inequality is finite ($x^y_{q,p}$ means that y_j occur on the places corresponding to the indices $j\in I\setminus I(q,p)$).

Corollary 3.75. *Under the assumptions of Theorem 3.73, the asymptotic relation (3.74) holds provided*

$$\int_{\mathbb{R}^d_+}|\gamma(y)|\,dy\le C_b\sum_{q=0}^{d-1}\sum_{p=1}^{\binom{d}{k}}\int_{\mathbb{R}_+}\left(\int_{\substack{y_j\le x_j,\\ j\in I(q,p)}}|(\partial_{q,p}f)_{s,r}(x^y_{q,p})|^b\prod_{j\in I(q,p)}\frac{dx_j}{y_j}\right)^{\frac1b}\times\prod_{j\in I\setminus I(q,p)}\frac{\cos 2\pi a_j}{y_j}\,dy$$

is finite for some $b>1$.

Corollary 3.76. *Under the assumptions of Theorem 3.73, the asymptotic relation (3.74) holds provided*

$$\int_{\mathbb{R}^d_+}|\gamma(y)|\,dy\le C\sum_{q=0}^{d-1}\sum_{p=1}^{\binom{d}{k}}\int_{\mathbb{R}_+}\operatorname*{ess\,sup}_{\substack{y_j\le x_j,\\ j\in I(q,p)}}|(\partial_{q,p}f)_{s,r}(x^y_{q,p})|\prod_{j\in I\setminus I(q,p)}\frac{\cos 2\pi a_j}{y_j}\,dy<\infty.$$

We do not give details of the proofs, just repeat that we provide one-dimensional techniques in each variable – proofs rather than results. Nothing but some accuracy in notation is needed for this. However, one more issue should be observed.

The finiteness of the bounds in the above theorem and corollaries also means that the considered functions are of (certain) bounded variation. There are many notions of bounded variation in the multidimensional case; for references and further results, see a recent paper [134].

One of the most natural multidimensional variations is Hardy's variation, which in fact is used in the above result. The following criterion for f to be of bounded Hardy variation can be found, for example, in [1, Th.6] (in dimension two).

Let $\eta = (\eta_1, \dots, \eta_d)$ be a d-dimensional vector with the entries either 0 or 1 only. Correspondingly, $|\eta| = \eta_1 + \cdots + \eta_d$. The inequality of vectors is meant coordinate wise. Denote by $\Delta_{u_\eta} f(x)$ the partial difference

$$\Delta_{u_\eta} f(x) = \Big(\prod_{j:\eta_j=1} \Delta_{u_j} \Big) f(x).$$

Here and in what follows $\partial^\eta f$ for $\eta = \mathbf{0} = (0, 0, \dots, 0)$ or $\eta = \mathbf{1} = (1, 1, \dots, 1)$ mean the function itself and the partial derivative repeatedly in each variable, respectively, where

$$\partial^\eta f(x) = \Big(\prod_{j:\,\eta_j=1} \frac{\partial}{\partial x_j} \Big) f(x).$$

Theorem 3.77. *A necessary and sufficient condition that $f(x)$ be of bounded Hardy variation is that it be expressible as the difference between two bounded functions, $f_1(x)$ and $f_2(x)$, satisfying the inequalities* $(i = 1, 2)$

$$\Delta_{u_\eta} f_i(x) \ge 0 \tag{3.78}$$

for all $\eta \ne \mathbf{0}$.

It is mentioned in the same paper [1] that such functions $f_1(x)$ and $f_2(x)$, which definitely express certain monotonicity property have been called "monotonely monotone" in [212]. They belong to the class of "quasimonotone" functions as defined in [97, p. 347]. The latter name is commonly used today for a completely different class of functions.

As in dimension one and following Bochner [31], we will mainly restrict ourselves to functions with Lebesgue integrable derivatives, since every such function is equivalent to a function of bounded variation in the sense that it is representable as a linear combination (generally, with complex coefficients) of monotone functions.

In several dimensions we apply a construction similar to that in dimension one (3.36). We consider a function f to be with each derivative $\partial^\eta f$, $\eta \ne \mathbf{0}$, existing almost everywhere and Lebesgue integrable with respect to dx_η for almost every value of the rest of the variables. The d-dimensional analog of the above-mentioned representation is as follows. Let $f(x)$ be defined on a rectangle

$[a, b] = [a_1, b_1] \times \cdots \times [a_d, b_d]$. Then

$$f(x) = f(b) + \sum_{\eta \neq \mathbf{0}} (-1)^{|\eta|} \Bigg[\int\limits_{[x_\eta, b_\eta]} \frac{|\partial^\eta f(u\eta, b_{\mathbf{1}-\eta})| + \partial^\eta f(u\eta, b_{\mathbf{1}-\eta})}{2} \, du\eta - \int\limits_{[x_\eta, b_\eta]} \frac{|\partial^\eta f(u\eta, b_{\mathbf{1}-\eta})| - \partial^\eta f(u\eta, b_{\mathbf{1}-\eta})}{2} \, du\eta \Bigg].$$

For example, in dimension two, for a function $f(x_1, x_2)$ on the rectangle $[a_1, b_1] \times [a_2, b_2]$, we have

$$\begin{aligned} f(x_1, x_2) = f(b_1, b_2) &+ \int_{x_1}^{b_1} \frac{|\partial^{(1,0)} f(t_1, b_2)| - \partial^{(1,0)} f(t_1, b_2)}{2} \, dt_1 \\ &- \int_{x_1}^{b_1} \frac{|\partial^{(1,0)} f(t_1, b_2)| + \partial^{(1,0)} f(t_1, b_2)}{2} \, dt_1 \\ &+ \int_{x_2}^{b_2} \frac{|\partial^{(0,1)} f(b_1, t_2)| - \partial^{(0,1)} f(b_1, t_2)}{2} \, dt_2 \\ &- \int_{x_2}^{b_2} \frac{|\partial^{(0,1)} f(b_1, t_2)| + \partial^{(0,1)} f(b_1, t_2)}{2} \, dt_2 \\ &- \int_{x_1}^{b_1} \int_{x_2}^{b_2} \frac{|\partial^{(1,1)} f(t_1, t_2)| - \partial^{(1,1)} f(t_1, t_2)}{2} \, dt_1 \, dt_2 \\ &+ \int_{x_1}^{b_1} \int_{x_2}^{b_2} \frac{|\partial^{(1,1)} f(t_1, t_2)| + \partial^{(1,1)} f(t_1, t_2)}{2} \, dt_1 \, dt_2. \end{aligned}$$

In dimension three, for a function $f(x_1, x_2, x_3)$ on the rectangle $[a_1, b_1] \times [a_2, b_2] \times [a_3, b_3]$, we have

$$\begin{aligned} f(x_1, x_2, x_3) = f(b_1, b_2, b_3) &+ \int_{x_1}^{b_1} \frac{|\partial^{(1,0,0)} f(t_1, b_2, b_3)| - \partial^{(1,0,0)} f(t_1, b_2, b_3)}{2} \, dt_1 \\ &- \int_{x_1}^{b_1} \frac{|\partial^{(1,0,0)} f(t_1, b_2, b_3)| + \partial^{(1,0,0)} f(t_1, b_2, b_3)}{2} \, dt_1 \\ &+ \int_{x_2}^{b_2} \frac{|\partial^{(0,1,0)} f(b_1, t_2, b_3)| - \partial^{(0,1,0)} f(b_1, t_2, b_3)}{2} \, dt_2 \end{aligned}$$

$$
\begin{aligned}
&- \int_{x_2}^{b_2} \frac{|\partial^{(0,1,0)} f(b_1,t_2,b_3)| + \partial^{(0,1,0)} f(b_1,t_2,b_3)}{2}\, dt_2 \\
&+ \int_{x_3}^{b_3} \frac{|\partial^{(0,0,1)} f(b_1,b_2,t_3)| - \partial^{(0,0,1)} f(b_1,b_2,t_3)}{2}\, dt_3 \\
&- \int_{x_3}^{b_3} \frac{|\partial^{(0,0,1)} f(b_1,b_2,t_3)| + \partial^{(0,0,1)} f(b_1,b_2,t_3)}{2}\, dt_3 \\
&- \int_{x_1}^{b_1}\int_{x_2}^{b_2} \frac{|\partial^{(1,1,0)} f(t_1,t_2,b_3)| - \partial^{(1,1,0)} f(t_1,t_2,b_3)}{2}\, dt_1\, dt_2 \\
&+ \int_{x_1}^{b_1}\int_{x_2}^{b_2} \frac{|\partial^{(1,1,0)} f(t_1,t_2,b_3)| + \partial^{(1,1,0)} f(t_1,t_2,b_3)}{2}\, dt_1\, dt_2 \\
&- \int_{x_1}^{b_1}\int_{x_3}^{b_3} \frac{|\partial^{(1,0,1)} f(t_1,b_2,t_3)| - \partial^{(1,0,1)} f(t_1,b_2,t_3)}{2}\, dt_1\, dt_3 \\
&+ \int_{x_1}^{b_1}\int_{x_3}^{b_3} \frac{|\partial^{(1,0,1)} f(t_1,b_2,t_3)| + \partial^{(1,0,1)} f(t_1,b_2,t_3)}{2}\, dt_1\, dt_3 \\
&- \int_{x_2}^{b_2}\int_{x_3}^{b_3} \frac{|\partial^{(0,1,1)} f(b_1,t_2,t_3)| - \partial^{(0,1,1)} f(b_1,t_2,t_3)}{2}\, dt_2\, dt_3 \\
&+ \int_{x_2}^{b_2}\int_{x_3}^{b_3} \frac{|\partial^{(0,1,1)} f(b_1,t_2,t_3)| + \partial^{(0,1,1)} f(b_1,t_2,t_3)}{2}\, dt_2\, dt_3 \\
&+ \int_{x_1}^{b_1}\int_{x_2}^{b_2}\int_{x_3}^{b_3} \frac{|\partial^{(1,1,1)} f(t_1,t_2,t_3)| - \partial^{(1,1,1)} f(t_1,t_2,t_3)}{2}\, dt_1\, dt_2\, dt_3 \\
&- \int_{x_1}^{b_1}\int_{x_2}^{b_2}\int_{x_3}^{b_3} \frac{|\partial^{(1,1,1)} f(t_1,t_2,t_3)| + \partial^{(1,1,1)} f(t_1,t_2,t_3)}{2}\, dt_1\, dt_2\, dt_3.
\end{aligned}
$$

All the integrals with $+$ before them form f_1, while those preceded with $-$ form f_2, for which one can easily check the sufficient conditions of Theorem 3.77.

By analogy with the one-dimensional case, applications to trigonometric series are in order. However, instead of giving explicitly estimates for trigonometric series similar to those in dimension one, moreover they are given in [125] along with a multivariate analog of (3.68), we prefer a different, two-dimensional, illustration of how such theorems work.

Let $f(x_1,x_2)$ be defined on $\mathbb{R}^2_+$. We will use the notation

$$f_1 := \frac{\partial f}{\partial x_1}, \quad f_2 := \frac{\partial f}{\partial x_2}, \quad \text{and} \quad f_{12} := \frac{\partial^2 f}{\partial x_1 \partial x_2}.$$

Next, for g defined on $\mathbb{R}^2_+$ let $g(x_1^*,x_2)$ mean that g is extended in the odd way in the first variable from $[0,\infty)$ to the whole $\mathbb{R}$. Similarly $g(x_1,x_2^*)$ means that g is extended in the odd way in the second variable from $[0,\infty)$ to the whole $\mathbb{R}$, and $g(x_1^*,x_2^*)$ means that g is extended in the odd way in each variable from $[0,\infty)$ to the whole $\mathbb{R}$.

Let f be locally absolutely continuous in each variable for any value of the other one, while $f_1(x_1,x_2)$ and $f_2(x_1,x_2)$ be locally absolutely continuous in x_2 and x_1 with respect to any value of x_1 and x_2, respectively, and let

$$\lim_{x_1+x_2\to\infty} f(x_1,x_2), f_1(x_1,x_2), f_2(x_1,x_2) = 0.$$

It is quite possible that these conditions are not minimal, that is, some may follow from the others (see, *e.g.*, [83] where absolute (not local) continuity – in both variables and in each variable for any value of the other one – is assumed along with $\lim\limits_{x_1+x_2\to\infty} f(x_1,x_2) = 0$ only, the derivatives vanish at infinity automatically then).

We now formulate a two-dimensional version of Corollary 3.76.

Corollary 3.79. *For any $y_1,y_2>0$ and $a_j = 0$ or $\frac14$, $j=1,2$,*

$$\int_{\mathbb{R}^2_+} f(x_1,x_2)\sin 2\pi(y_1x_1+a_1)\sin 2\pi(y_2x_2+a_2)\,dx_1dx_2$$
$$= \frac{1-4a_1}{2\pi y_1}\frac{1-4a_2}{2\pi y_2} f\left(\frac{1}{4y_1},\frac{1}{4y_2}\right) + \phi(y_1,y_2),$$

where

$$\int_{\mathbb{R}^2_+} |\phi(y_1,y_2)|\,dy_1dy_2 \le C\int_{\mathbb{R}^2_+} \operatorname*{ess\,sup}_{\substack{y_1\le x_1\le 2y_1,\\ y_2\le x_2\le 2y_2}} |f_{12}(x_1,x_2)|\,dy_1dy_2$$
$$+C\int_{\mathbb{R}^2_+} \operatorname*{ess\,sup}_{y_1\le x_1\le 2y_1} |f_1(x_1,y_2)|\frac{\cos 2\pi a_2}{y_2}\,dy_1dy_2$$
$$+C\int_{\mathbb{R}^2_+} \operatorname*{ess\,sup}_{y_2\le x_2\le 2y_2} |f_2(y_1,x_2)|\frac{\cos 2\pi a_1}{y_1}\,dy_1dy_2 < \infty.$$

Now, in [21] a result on the integrability of the Fourier transform is given; the next is its two-dimensional version.

Theorem 3.80. *If the function f satisfies the above absolute continuity assumptions and for some $q \in (0,1)$*

$$|f(x_1,x_2)| \le C(1+|x_1|)^{-q}(1+|x_2|)^{-q}, \tag{3.81}$$
$$|f_1(x_1,x_2)| \le C(1+|x_1|)^{-2q}(1+|x_2|)^{-q}|x_1|^{q-1}, \tag{3.82}$$
$$|f_2(x_1,x_2)| \le C(1+|x_1|)^{-q}(1+|x_2|)^{-2q}|x_2|^{q-1}, \tag{3.83}$$

and

$$|f_{12}(x_1,x_2)| \le C(1+|x_1|)^{-2q}(1+|x_2|)^{-2q}|x_1x_2|^{q-1}, \tag{3.84}$$

then the Fourier transform $\hat f$ is integrable.

The proof is straightforward but pretty long and not that easy. We will give a short proof that readily follows from Corollary 3.79.

Proof. There are three typical situations: both variables y_1 and y_2 are small, both are large, and one is small while the other is large.

We first represent the function as a sum of four functions from which one f_{ee} is even in each variable, two f_{eo} and f_{oe} are even in one variable and odd in the other, and the last f_{oo} is odd in each variable. We have

$$4f_{ee}(x_1,x_2) = f(x_1,x_2) + f(-x_1,x_2) + f(x_1,-x_2) + f(-x_1,-x_2),$$
$$4f_{eo}(x_1,x_2) = f(x_1,x_2) - f(x_1,-x_2) + f(-x_1,x_2) - f(-x_1,-x_2),$$
$$4f_{oe}(x_1,x_2) = f(x_1,x_2) - f(-x_1,x_2) + f(x_1,-x_2) - f(-x_1,-x_2),$$

and

$$4f_{oo}(x_1,x_2) = f(x_1,x_2) - f(-x_1,x_2) - f(x_1,-x_2) + f(-x_1,-x_2),$$

with $f = f_{ee} + f_{eo} + f_{oe} + f_{oo}$.

We then apply to each of these functions Corollary 3.79 and (3.81)–(3.84). The procedure goes smoothly and will be illustrated for the leading term of the asymptotics which exists only for f_{oo}. Integration over $y_1, y_2 > 1$ reduces to

$$\int_0^1 \int_0^1 \frac{1}{y_1}\frac{1}{y_2} \int_{-y_1}^{y_1} \int_{-y_2}^{y_2} |f_{12}(u_1,u_2)\, du_1 du_2|\, dy_1 dy_2,$$

and (3.84) along with obvious estimates gives the convergence. For $y_1, y_2 < 1$ we obtain

$$\int_1^\infty \int_1^\infty \frac{1}{y_1}\frac{1}{y_2} |f_{oo}(y_1,y_2)|\, dy_1 dy_2,$$

and (3.81) immediately yields the needed bound. Let now $y_1 < 1$ and $y_2 > 1$. We have

$$\int_0^1 \int_1^\infty \frac{1}{y_1}\frac{1}{y_2} |f_{oo}(y_1,y_2)|\, dy_1 dy_2$$
$$= \int_1^\infty \frac{1}{y_2} \int_0^1 \frac{1}{y_1}\frac{1}{y_2} \int_{-y_1}^{y_1} |f_1(u_1,y_2) - f_1(u_1,-y_2)|\, du_1\, dy_1\, dy_2,$$

and (3.82) leads to evident estimates. In the "mixed" terms where we have ess sup in one variable and y^{-1} in the other, we just estimate the latter by splitting the integral into two over $[0,1]$ and $[1,\infty]$. The procedure is exactly the same as the one for the "mixed" integration in the leading term. The same is for f_{eo} and f_{oe}.

The proof is complete. □

This proof also shows that Theorem 3.73 as well as its corollaries are sharp both at the origin and at infinity.

Of course, a more subtle version of Theorem 3.80 can be given, with logarithmic conditions say, and of course the same immediate proof rather than a long technical straightforward one can be given by means of Corollary 3.79.

Chapter 4

The Fourier Transform of a Radial Function

Spherical symmetry is a very interesting and important property of a function. Theorem 1.5 gives that if $f(x)$ is radial (depending only on $|x|$), then $\widehat{f}$ is radial too. More precisely, it is represented as a Hankel transform as the Cauchy–Poisson formula (1.7), by Theorem 1.6. For radial L^p functions this formula makes sense for $1 < p < (2d/(d+1))$.

Note that the restriction of the multiplier problem for the ball to radial functions has a positive solution for $|(1/p) - (1/2)| \geq (1/2d)$; see [95].

Though, generally speaking, the Fourier transform should be understood in the distributional sense, we are going to consider a situation where the Fourier transform of a non-integrable radial function defined in the distributional sense turns out to be a regular function represented via an integral similar to that in Theorem 1.6. In this setting the space of functions (of one variable) of bounded variation is generalized, and the Fourier transform exists as an improper integral (see, *e.g.*, [30]).

The property of a function to be radial combined with the distributional approach gives the following result (see [146]).

Theorem 4.1. *Let f grow at infinity not faster than a polynomial. Then*

$$\widehat{f}(\xi) = \lim_{A\to\infty} 2\pi \int_0^A f_0(t)(|\xi|t)^{1-\frac{d}{2}} J_{\frac{d}{2}-1}(2\pi|\xi|t)t^{d-1}dt,$$

where convergence is that in topology of $\mathcal{S}'$, the weak convergence.

Proof. For the considered functions f, we have $f_A \to f_0$ in the topology of the space of distributions, where $f_A(t) = f_0(t)$ if $t \leq A$ and zero otherwise. This completes the proof, since to each (integrable) function f_A we can apply Theorem 1.6. Observe that in general $\widehat{f}(\xi)$ is a distribution. $\square$

This will be our starting point in treating the Fourier transform but first we must define an appropriate class of functions.

4.1 Fractional derivatives and classes of functions

We first need to dwell upon a notion of fractional derivative. For $0 < \delta < 1$ and a locally integrable function g on $(0, \infty)$, define the fractional (Weyl type) integral of order δ by

$$W_\omega^\delta g(t) = \begin{cases} \frac{1}{\Gamma(\delta)} \int_t^\omega g(r)(r-t)^{\delta-1} dr, & 0 < t < \omega, \\ 0, & t \geq \omega, \end{cases}$$

and, following J. Cossar [59], a fractional Weyl derivative of order α by

$$g^{(\alpha)}(t) = \lim_{\omega \to \infty} -\frac{d}{dt} W_\omega^{1-\alpha} g(t)$$

when $0 < \alpha < 1$ and

$$g^{(\alpha)}(t) = \frac{d^p}{dt^p} g^{(\delta)}(t)$$

when $\alpha = p + \delta$ with $p = 1, 2, \ldots$, and $0 < \delta < 1$. One of the reasons that just this type of fractional integral (and derivative) is chosen is that the Weyl integral of a function with compact support has, in turn, compact support, unlike the better known Riemann–Liouville integral

$$R_\alpha(f_0; t) = \frac{1}{\Gamma(\alpha)} \int_0^t f_0(r)\, (t-r)^{\alpha-1} dr.$$

All these notions may be found, for example, in [11, Ch. 13] (see also [167] and [194]).

Denote by AC_{loc} and BV_{loc} the classes of functions locally absolutely continuous and locally of bounded variation, respectively. Let α^* be the greatest integer less than α. If α is fractional $\alpha^* = [\alpha]$, where $[\cdot]$ denotes the integer part, while for α integer $\alpha^* = \alpha - 1$.

Consider the class $MV_{\alpha+1}^b$, with $\alpha > 0$ and $b \geq 0$, of $C(0,\infty)$-functions satisfying the following conditions

$$g, g', \ldots, g^{(\alpha^*)} \quad \text{are} \quad AC_{\text{loc}} \quad \text{on} \quad (0, \infty); \tag{4.2}$$

$$\lim_{t \to \infty} g(t), \quad \lim_{t \to \infty} t^{\alpha+b} g^{(\alpha)}(t) = 0; \tag{4.3}$$

and

$$\|g\|_{MV_{\alpha+1}^b} := \sup_{t>0} |t^b g(t)| + \int_0^\infty \left| d[(t^{\alpha+b} g^{(\alpha)}(t)] \right| < \infty. \tag{4.4}$$

It is of considerable interest to compare this space with a related one, written $BV_{\alpha+1}^b$, proven to be useful in problems of approximation and multipliers (see, *e.g.*, [48] or [194]). It is defined to be the space of $C(0,\infty)$-functions satisfying the following conditions:

$$g^{(\alpha-[\alpha])},\dots,g^{(\alpha-1)} \quad \text{are} \quad AC_{\mathrm{loc}} \quad \text{on} \quad (0,\infty); \tag{4.5}$$

$$\lim_{t\to\infty} g(t)=0, \quad g^{(\alpha)}\in BV_{\mathrm{loc}}(0,\infty); \tag{4.6}$$

and

$$\|g\|_{BV^b_{\alpha+1}} := \int_0^\infty t^{\alpha+b}|dg^{(\alpha)}(t)|<\infty. \tag{4.7}$$

The main difference, of course, comes from the definitions of the norms (4.4) and (4.7), with the factor $t^{\alpha+b}$ within or beyond the sign of the differential, respectively. This is hardly expected to be meaningful. Nevertheless, the continuous embedding

$$BV^b_{\alpha+1}\hookrightarrow MV^b_{\alpha+1}$$

holds, and the example which demonstrates the difference is delivered by a strongly oscillating function (see below).

For completeness, we compare these spaces with the other space $WBV^b_{\infty,\alpha+1}$, as considered, *e.g.*, in [79, 80], of $C(0,\infty)$ functions satisfying (4.5) and additionally $g^{(\alpha)}\in AC_{\mathrm{loc}}(0,\infty)$, and

$$\|g\|_{WBV^b_{\infty,\alpha+1}} := \operatorname*{ess\,sup}_{t>0}|t^b g(t)| + \operatorname*{ess\,sup}_{t>0}|t^{\alpha+1+b}g^{(\alpha+1)}(t)|<\infty. \tag{4.8}$$

Various conditions for Fourier multipliers were given in terms of these classes (see references above). The formulas by means of which these spaces are defined have proved to be convenient to express multipliers in explicit form.

The following assertion establishes relations between all these classes.

Proposition 4.9. *In the sense of continuous embedding there holds*

$$BV^b_{\alpha+1}\hookrightarrow MV^b_{\alpha+1}\hookrightarrow WBV^b_{\infty,\alpha}.$$

None of these embeddings holds in the opposite direction.

Proof. We only discuss the instance $b=0$ which is the worst case. Here and in what follows we omit the superscript b in the case $b=0$. Suppose $g\in BV_{\alpha+1}$; then, by Lemma 1.1 in [195],

$$\int_0^\infty |d(t^\alpha g^{(\alpha)}(t))| \le C\int_0^\infty t^{\alpha-1}|g^{(\alpha)}(t)|\,dt + C\int_0^\infty t^\alpha|dg^{(\alpha)}(t)| \le C\|g\|_{BV_{\alpha+1}},$$

showing the norm estimate for the left inclusion. Trivially, a function h of bounded variation, vanishing at infinity, satisfies

$$|h(t)|\le\int_0^\infty |dh(t)|,$$

hence $\|g\|_{WBV_{\infty,\alpha}} \le \|g\|_{MV_{\alpha+1}}$.

The absolute continuity properties follow from the discussion in Section 3 of [79]; further, by Lemma 3.15 in [194],

$$|t^\alpha g^{(\alpha)}(t)| = Ct^\alpha \left| \int_t^\infty dg^{(\alpha)}(s) \right| \le C \int_t^\infty s^\alpha |dg^{(\alpha)}(s)|,$$

with $g \in BV_{\alpha+1}$, which, by hypothesis, tends to zero as $t \to \infty$ a.e., so the embeddings are proved.

That the inclusions are strict, show the following examples ($b = 0$). Let first

$$g_1(t) = \chi(t) \sin\left(\ln\ln\left(\frac{1}{t}\right)\right), \tag{4.10}$$

where $\chi \in C^\infty[0,\infty)$ is a cut-off function which is 1 for $0 \le t \le \frac{1}{4}$ and vanishes for $t \ge \frac{1}{2}$. Then

$$g_1'(t) = -\frac{\cos(\ln\ln(\frac{1}{t}))}{t\ln(\frac{1}{t})}$$

for $0 < t < \frac{1}{4}$ and $g_1'(t) = 0$ for $t > \frac{1}{2}$, hence $tg_1'(t)$ is of bounded variation on $[0,\infty)$ but

$$\begin{aligned}\|g_1\|_{BV_1} &\ge \int_0^{\frac{1}{4}} |g_1'(t)|\,dt = \int_{\ln 4}^\infty |\cos(\ln u)| \frac{du}{u} \\ &\ge \int_1^\infty |\cos v|\,dv = \infty,\end{aligned}$$

thus $g_1 \notin BV_2$ since $\|g_1\|_{BV_1} \le C\|g_1\|_{BV_2}$. This argument also works for all $\alpha > 0$, $b = 0$, if one replaces g_1 by

$$g_{1,\alpha}(t) = \int_t^1 (u-t)^{\delta-1} \frac{\cos(\ln\ln(\frac{1}{u}))}{u^\delta \ln(\frac{1}{u})} \chi(u)\,du, \tag{4.11}$$

with $\alpha = p + \delta$, $0 < \delta \le 1$; note that $g_{1,\alpha}$ is bounded and that

$$g_{1,\alpha}^{(\delta)}(t) = C\frac{\cos(\ln\ln(\frac{1}{t}))}{t^\delta \ln(\frac{1}{t})}\chi(t).$$

In the case $\alpha = 1$, (4.10) is an extension of the earlier suggestion by Belinsky (see [120]) to consider

$$g_2(t) = \frac{g_1(t)}{\ln\ln(\frac{1}{t})}$$

which also delivers an example of the strict embedding; observe that g_2 is continuous at the origin and vanishes there in contrast to (4.10).

Similarly consider $g_3(t) = t^{i\gamma}$, with $\gamma \in \mathbb{R}$; then $g_3 \in WBV_{\infty,\alpha}$ for all $\alpha \ge 0$ and obviously $t^{i\gamma} \notin MV_{\alpha+1}$ for all $\alpha > 0$ and $\gamma \in \mathbb{R}$ fixed. □

There are another two important properties of functions from the $MV_{\alpha+1}^b$ classes.

Lemma 4.12. *If $g \in MV_{\alpha+1}^b$, then it follows for $p = 1, 2, \ldots, \alpha^*$ that*

$$\|t^{b+p} g^{(p)}(t)\|_\infty \le C\|g\|_{MV_{\alpha+1}^b} \qquad \text{and} \qquad \lim_{t\to\infty} t^p g^{(p)}(t) = 0.$$

Proof. The first assertion is clear on account of the embedding behavior of the WBV-spaces: $WBV_{\infty,\alpha}^b \hookrightarrow WBV_{\infty,\beta}^b$ for $0 < \beta < \alpha$. From this also the second assertion is obvious in the case $b > 0$. So let $b = 0$ and first suppose α is integer. Then the argument in [206, p. 193] shows the assertion in the case $\alpha = 2$. That this argument also works for all α integer we illustrate at the case $\alpha = 3$. Thus we show: $\lim_{t\to\infty} g(t) = 0$ and $\lim_{t\to\infty} t^3 g'''(t) = 0$ implies $\lim_{t\to\infty} t^2 g''(t) = 0$ whence follows $\lim_{t\to\infty} t g'(t) = 0$ by the case $\alpha = 2$. Let $0 < \delta < \frac{1}{4}$ be arbitrary, fixed, $t > 0$ be large. Then by Taylor's formula (at $s = 0$)

$$\begin{aligned} G(s) &:= \Delta_s^2 g(t) = g(t+2s) - 2g(t+s) + f(t) \\ &= g''(t)s^2 + (s^3/3!)\Big(8g'''(t+2\theta s) - 2g'''(t+\theta s)\Big) \end{aligned}$$

with $0 < \theta < 1$. Now choose $s = \pm\delta t$ and T_δ so large that $|\Delta_{\delta t}^2 g(t)| \le \delta^3$ for all $t \ge T_\delta$. Furthermore, observe that by hypothesis

$$|G'''(\theta s)| \le 8M(t + 2\theta\delta t)^{-3} + 2M(t + \theta\delta t)^{-3} \le 10Mt^{-3}(1-2\delta)^{-3}$$

and thus

$$|t^2 g''(t)| = \left|\delta^{-2}\Delta_{\pm\delta t}^2 g(t) - \frac{\delta t^3 G'''(\theta s)}{3!}\right| \le \delta\left(1 + \frac{10M(1-2\delta)^{-3}}{3!}\right)$$

for $t \ge T_\delta$, *i.e.*, the assertion for $\alpha = 3$. It is clear that by using higher differences, the remaining case of α integer can be proved by induction.

If α is fractional and $\alpha > 1$, then we have, by Lemma 1, that

$$g \in WBV_{\infty,[\alpha]} \hookrightarrow WBV_{\infty,j}, \quad j = 1, \ldots, [\alpha],$$

and the argument for α integer shows $\lim_{t\to\infty} t^j g^{(j)}(t) = 0$ for $j = 1, \ldots, [\alpha] - 1$. Thus we only have to look at the case $j = [\alpha] < \alpha$. By Proposition 4.9, we have $g \in WBV_{\infty,\alpha}$ and hence the fractional calculus within the WBV-spaces (see [79], [80]) may be applied to give with an integration by parts

$$\begin{aligned} g^{([\alpha])}(t) &= C\int_t^\infty (s-t)^{\alpha-[\alpha]-1} s^{-\alpha} s^\alpha g^{(\alpha)}(s)\, ds \\ &= C\left(\int_s^\infty (\sigma-t)^{\alpha-[\alpha]-1}\sigma^{-\alpha}\, d\sigma\right) s^\alpha g^{(\alpha)}(s)\,\Big|_{s=t}^{\infty} \\ &\quad + C\int_t^\infty \left(\int_s^\infty (\sigma-t)^{\alpha-[\alpha]-1}\sigma^{-\alpha}\, d\sigma\right) d\Big(s^\alpha g^{(\alpha)}(s)\Big) \\ &= I_t + II_t. \end{aligned}$$

With the substitution $\sigma = tr$, the term I_t can be estimated by ($s \ge t$, $\alpha > 1$)

$$\begin{aligned}|t^{[\alpha]} I_t| &\le C\Big(\int_{s/t}^{\infty} (r-1)^{\alpha-[\alpha]-1} r^{-\alpha}\,dr\Big) s^{\alpha} g^{(\alpha)}(s)\Big|_{s=t}^{\infty} \\ &= C\Big(\int_{1}^{\infty} (r-1)^{\alpha-[\alpha]-1} r^{-\alpha}\,dr\Big) t^{\alpha} g^{(\alpha)}(t),\end{aligned}$$

therefore $|t^{[\alpha]} I_t| \to 0$ as $t \to \infty$, since by hypothesis $\lim\limits_{t\to\infty} t^{\alpha} g^{(\alpha)}(t) = 0$.

Concerning II_t an interchange of integration gives

$$\begin{aligned}|II_t| &= C\int_t^{\infty} (\sigma - t)^{\alpha-[\alpha]-1} \sigma^{-\alpha} \int_t^{\sigma} |d(s^{\alpha} g^{(\alpha)}(s))|\,d\sigma \\ &\le C t^{-[\alpha]} \int_t^{\infty} |d(s^{\alpha} g^{(\alpha)}(s))| \int_1^{\infty} (r-1)^{\alpha-[\alpha]-1} r^{-\alpha}\,dr\end{aligned}$$

thus also $|t^{[\alpha]} II_t| \to 0$ as $t \to \infty$, since

$$\int_0^{\infty} |d(s^{\alpha} g^{(\alpha)}(s))| < \infty.$$

Combining these two situations gives the remaining case

$$\lim_{t\to\infty} t^{[\alpha]} g^{([\alpha])}(t) = 0,$$

which completes the proof. □

The next fact true for functions from $BV^b_{\alpha+1}$ (see this and much more details in [194]) turns out to be true for a wider class as well.

Lemma 4.13. *For $g \in MV^b_{\alpha+1}$, with $\alpha = p + \delta$, $p = 1, 2, \ldots$,*

$$g^{(\alpha)}(t) = (g^{(p)})^{(\delta)}(t)$$

for all $t \in (0, \infty)$.

Proof. Recall that

$$g^{(\delta)}(t) = \lim_{\omega\to\infty} -\frac{d}{dt} W_{\omega}^{1-\delta} g(t).$$

Integrating $W_{\omega}^{1-\delta}$ by parts p times, we obtain

$$\begin{aligned}W_{\omega}^{1-\delta} g(t) &= \frac{1}{\Gamma(1-\delta)} \sum_{k=1}^{p} \frac{(\omega - t)^{k-\delta}}{(1-\delta)\cdots(k-\delta)} (-1)^{k+1} g^{(k-1)}(\omega) \\ &\quad + (-1)^p \frac{1}{\Gamma(1-\delta)(1-\delta)\cdots(p-\delta)} \int_t^{\omega} (y-t)^{p-\delta} g^{(p)}(y)\,dy.\end{aligned}$$

Taking $-\frac{d}{dt}$ from the both sides, we obtain

$$-\frac{d}{dt}W_\omega^{1-\delta}g(t) = \frac{1}{\Gamma(1-\delta)}\sum_{k=1}^{p}\frac{(\omega-t)^{k-1-\delta}}{(1-\delta)\cdots(k-1-\delta)}(-1)^{k+1}g^{(k-1)}(\omega)$$
$$+(-1)^p\frac{1}{\Gamma(1-\delta)(1-\delta)\cdots(p-1-\delta)}\int_t^\omega (y-t)^{p-1-\delta}g^{(p)}(y)\,dy.$$

From Lemma 4.12 it follows that the sum on the right tends to zero as $\omega\to 0$. The limit as $\omega\to 0$ of the last integral exists since exists – by definition – the limit of the left-hand side. By this

$$g^{(\delta)}(t) = \frac{(-1)^p}{\Gamma(1-\delta)(1-\delta)\cdots(p-1-\delta)}\int_t^\infty (y-t)^{p-1-\delta}g^{(p)}(y)\,dy.$$

Differentiating both sides p times, we have

$$\frac{d^p}{dt^p}g^{(\delta)}(t) = \frac{d^p}{dt^p}\frac{(-1)^p}{\Gamma(1-\delta)(1-\delta)\cdots(p-1-\delta)}\int_t^\omega (y-t)^{p-1-\delta}g^{(p)}(y)\,dy$$
$$+\frac{d^p}{dt^p}\frac{(-1)^p}{\Gamma(1-\delta)(1-\delta)\cdots(p-1-\delta)}\int_\omega^\infty (y-t)^{p-1-\delta}g^{(p)}(y)\,dy$$
$$-\frac{d}{dt}\frac{1}{\Gamma(1-\delta)}\int_t^\omega (y-t)^{-\delta}g^{(p)}(y)\,dy$$
$$-\frac{d}{dt}\frac{1}{\Gamma(1-\delta)}\int_\omega^\infty (y-t)^{-\delta}g^{(p)}(y)\,dy.$$

To justify differentiation in the last integral we observe that the result

$$-\frac{\delta}{\Gamma(1-\delta)}\int_\omega^\infty (y-t)^{-1-\delta}g^{(p)}(y)\,dy$$

is uniformly bounded for t fixed and ω large enough as well as similar integrals for lower derivatives. Moreover, it obviously tends to zero as $\omega\to 0$. We then obtain

$$\frac{d^p}{dt^p}g^{(\delta)}(t) = \lim_{\omega\to 0} -\frac{d}{dt}\frac{1}{\Gamma(1-\delta)}\int_t^\omega (y-t)^{-\delta}g^{(p)}(y)\,dy = (g^{(p)})^{(\delta)}(t),$$

the desired relation. □

4.2 Existence of the Fourier transform and Bessel type functions

We concentrate on the spaces $MV_{\alpha+1}^b$ with

$$0<\alpha\le\frac{d-1}{2} \quad \text{and} \quad b=\frac{d-1}{2}-\alpha.$$

Set

$$F_\alpha(t) = t^{\frac{d-1}{2}} f_0^{(\alpha)}(t).$$

In the limiting case $\alpha = \frac{d-1}{2}$, $b = 0$, we write

$$F(t) := F_{\frac{d-1}{2}}(t), \quad \text{and} \quad \mathrm{MV} := \mathrm{MV}^0_{\frac{d+1}{2}}.$$

It turns out for just this range of α and b that, although a priori the Fourier transform should be understood in the distributional sense, it does exist in a regular sense.

Let, as usual, $C[a,b]$ and $C^p[a,b]$ be the classes of continuous functions and having p continuous derivatives, respectively, on $[a,b]$.

Let us introduce two Bessel-type functions:

$$\begin{aligned} Q_\alpha(t) &= \int_0^1 (1-s)^{\alpha-1} s^{\frac{d}{2}} J_{\frac{d}{2}-1}(ts)\, ds \\ &= \Gamma(\alpha) t^{-\frac{d}{2}-\alpha} R_\alpha(s^{\frac{d}{2}} J_{\frac{d}{2}-1}(s); t) \end{aligned}$$

and

$$q_\alpha(t) = \int_0^1 (1-s)^{\alpha-1} s^{\frac{d}{2}-1}\, J_{\frac{d}{2}}(ts) ds,$$

where J_μ is the Bessel function of first type and order μ. When $\alpha = \frac{d-1}{2}$, for brevity, we will write simply Q and q.

We denote by φ an arbitrary function $\varphi \in S$ and by φ_0 its radial part (spherical average).

The following result is the promised existence (and inversion) theorem.

Theorem 4.14. *Let f_0 be a function from $MV^b_{\alpha+1}$ with*

$$0 < \alpha \le \frac{d-1}{2}$$

and $b = \frac{d-1}{2} - \alpha$. Then there holds for the radial extension $f(x) = f_0(|x|)$ of f_0

$$\widehat{f}(u) = \frac{2\pi(-1)^{\alpha^*+1}}{\Gamma(\alpha)} |u|^{1-\frac{d}{2}} \int_0^\infty F_\alpha(t)\, t^{\alpha+\frac{1}{2}} Q_\alpha(2\pi|u|t)\, dt, \tag{4.15}$$

and $\widehat{f}(u)$ is continuous for $|u| > 0$, tends to zero as $|u| \to \infty$, and coincides with the distributional Fourier transform $\tilde{f}$ of f; for $|x| > 0$ the following inversion formula holds

$$f(x) = \lim_{A\to\infty} \int_{|u|\le A} \left(1 - \frac{|u|^2}{A^2}\right)^{\frac{d-1}{2}-\alpha} \widehat{f}(u)\, e^{2\pi i x\cdot u} du. \tag{4.16}$$

Both integrals converge uniformly away from the origin.

We note that Theorem 4.14 can handle radial functions which are not L^p-integrable, $p < \infty$; *e.g.*,

$$f_0 = \ln^{-1}(e + t), \quad \alpha = \frac{d-1}{2}.$$

Further, observe that the restriction $\alpha \le \frac{d-1}{2}$ is somehow natural, since $f \in L^1_{\mathrm{rad}}$ yields $\widehat{f_0} \in BV_{\gamma+1}$ with $\gamma < \frac{d-3}{2}$. Representations similar to (4.16) with

$$\alpha > d\left(\frac{1}{p} - \frac{1}{2}\right) - \frac{1}{2}$$

have been used earlier (see, *e.g.*, [195]), where one has the absolute convergence of the integrals involved. Theorem 4.14 is mainly based on handling improper Riemann integrals.

Let us compare Theorem 4.14 with another earlier results. Bochner in [35, §44] considers not only radial functions, but more restrictive conditions are claimed for the radial part of a function. Analogously, the radiality allows less restrictive smoothness conditions than those in V.A. Ilyin and Alimov theorems for general spectral expansions (see [105]). In Goldman's paper [87] radial functions are considered, with "worse" conditions at infinity and the monotonicity of a given function and its derivatives. The very simple Leray's formula somehow similar to (4.15) may be found in [167, Ch. 5, Lemma 25.1′], but sharp assumptions are still hidden in the fractional integral.

To prove the theorem we need auxiliary results for the introduced Bessel type functions, which are of interest by themselves.

Lemma 4.17. *The following asymptotic relation holds:*

$$q_\alpha(r) = \Gamma(\alpha)\, r^{-\alpha} J_{\frac{d}{2}+\alpha}(r) + \zeta_{\alpha,d} r^{-\frac{d}{2}} + O\big(r^{-\alpha-\frac{3}{2}}\big)$$

as $r \to \infty$, and $\zeta_{\alpha,d}$ is a number.

Proof. We have

$$q_\alpha(r) = \sum_{j=0}^{M} \int_0^1 (1-s)^{\alpha-1+j}(1+s)^j s^{\frac{d}{2}+1} J_{\frac{d}{2}}(rs)\, ds + \int_0^1 g_0(s)\, s^{\frac{d}{2}-1}\, J_{\frac{d}{2}}(rs)\, ds,$$

where M is such that

$$g_0(s) = (1-s)^{\alpha+M}(1+s)^{M+1}$$

is smooth enough at $s = 1$. Evaluate first the last integral. We need the following properties of the Bessel functions (see, *e.g.*, [11, §7.2.8 (50), (51); §7.13.1 (3);

§7.12 (8)]):

$$\frac{d}{dt}\,\left[t^{\pm\nu}J_\nu(t)\right] = \pm t^{\pm\nu}\,J_{\nu\mp1}(t); \tag{4.18}$$

$$J_\nu(t) = \sqrt{\frac{2}{\pi t}}\cos\left(t-\frac{\pi\nu}{2}-\frac{\pi}{4}\right)+\sqrt{\frac{2}{\pi}}\,\frac{1-4\nu^2}{8}\,t^{-\frac{3}{2}}\sin\left(t-\frac{\pi\nu}{2}-\frac{\pi}{4}\right) + O\big(t^{-\frac{5}{2}}\big) \quad \text{as} \quad t\to\infty; \tag{4.19}$$

$$J_\nu = O(t^\nu) \qquad \text{for small} \quad t. \tag{4.20}$$

Let us integrate by parts, using (4.18) as follows:

$$\begin{aligned}
&\int_0^1 g_0(s)\,s^{\frac{d}{2}-1}\,J_{\frac{d}{2}}(rs)\,ds\\
&= \int_0^1 [g_0(s)\,s^{d-2}]\,[s^{-\frac{d}{2}+1}\,J_{\frac{d}{2}}(rs)]\,ds\\
&= -\frac{1}{r}g_0(s)\,s^{\frac{d}{2}-1}\,J_{\frac{d}{2}-1}(rs)\,\Big|_0^1 + \frac{1}{r}\int_0^1 g_1(s)\,s^{\frac{d}{2}-2}\,J_{\frac{d}{2}-1}(rs)\,ds\\
&= \frac{1}{r}\int_0^1 g_1(s)\,s^{\frac{d}{2}-2}\,J_{\frac{d}{2}-1}(rs)\,ds,
\end{aligned}$$

where g_1 is also smooth enough. For the case $d=2$, unlike that for $d>2$, the factor $s^{\frac{d}{2}-2}$ does not appear in the last integral. For higher dimensions, we can continue this procedure, since the integrated terms may not vanish at $s=0$ only on the last step. After $[\frac{d}{2}]$ steps we get

$$\zeta_{\alpha,d}r^{-\frac{d}{2}} + r^{-\frac{d}{2}}\int_0^1 g_{[\frac{d}{2}]}(s)J_0(rs)\,ds = \zeta_{\alpha,d}r^{-\frac{d}{2}} + O\big(r^{-\frac{d}{2}-1}\big)$$

when d is even, and the integral

$$r^{\frac{1-d}{2}}\int_0^1 g_{[\frac{d}{2}]}(s)s^{-\frac{1}{2}}J_{\frac{1}{2}}(rs)\,ds = \zeta_{\alpha,d}r^{-\frac{d}{2}} + O\big(r^{-\frac{d}{2}-1}\big)$$

when d is odd. Clearly, $g_{[\frac{d}{2}]}$ denotes a relatively smooth function, like g_1 above.

To estimate the sum we need the following lemma.

Lemma 4.21. *For $r\ge 1$, $\beta > -\frac{1}{2}$, $\mu > -1$ and each positive integer p*

$$\int_0^1 (1-s)^\mu\,s^{\beta+1}J_\beta(rs)\,ds = \sum_{j=1}^{p}\alpha_j^\mu r^{-(\mu+j)}\,J_{\beta+\mu+j}(r) + O\big(r^{-\mu-p-\frac{3}{2}}\big),$$

where α_j^μ are some numbers depending only on j and μ, and

$$\alpha_1^\mu = \Gamma(\mu+1),\ \alpha_2^\mu = \mu\Gamma(\mu+2).$$

Proof. We have

$$(1-s)^\mu = 2^{-\mu}(1-s^2)^\mu + \left[(1-s)^\mu - 2^{-\mu}(1-s^2)^\mu\right]$$
$$= 2^{-\mu}(1-s^2)^\mu + 2^{-\mu}(1-s)^{\mu+1}\,\frac{2^\mu-(1+s)^\mu}{1-s}.$$

Continuing this process of chipping off the binomials $(1-s^2)^{\mu+j-1}$ for $j=1,\dots,p$, we use the formula (see, *e.g.*, [181, Ch. 4, Lemma 4.13]):

$$J_{\beta+\mu+1}(r) = \frac{r^{\mu+j}}{2^{\mu+j-1}\Gamma(\mu+j)}\int_0^1 J_\beta(rs)\; s^{\beta+1}(1-s^2)^{\mu+j-1}ds.$$

The remainder term is estimated as above by integrating by parts μ^*+p+1 times. Estimates are better in this case, since here s is in a rather high power. The lemma is proved. □

To finish the proof of Lemma 4.17, it remains to apply Lemma 4.21, with $\beta=\frac{d}{2}$, $\mu=\alpha-1+j$, and $p=1$, to the integrals in the sum for $q_\alpha(r)$. □

Remark 4.22. Sometimes the rough estimate

$$q_\alpha(r) = O\big(r^{-\alpha-\frac12}\big) \tag{4.23}$$

will be enough.

The following lemma is due to Trigub (see [199, Lemma 2]). The lemma concerns the functions

$$i(\mu,\lambda,r) = \int_0^1 t^\mu J_\lambda(rt)\;dt$$

where $\mu+\lambda>-1$.

Lemma 4.24. *There holds:*

1) $i(\mu,\lambda,r)=\frac1r\,J_{\lambda+1}(r)+\frac{\lambda+1-\mu}{r}\,i(\mu-1,\lambda+1,r)$.
2) *The function* $i(\mu,\lambda,r)$ *is* $O(r^\lambda)$ *for small* r, *and when* $r\to\infty$ *it behaves either as* $O(r^{-\frac32})$ *or* $O(r^{-1-\mu})$ *for* $\mu>\frac12$ *and* $\mu\le\frac12$, *respectively.*

Proof. To prove **1)**, we integrate by parts using (4.18). We have

$$i(\mu,\lambda,r) = \frac1r\int_0^1 t^{\mu-\lambda-1}d\left[t^{\lambda+1}J_{\lambda+1}(rt)\right]$$
$$= \frac1r J_{\lambda+1}(r) + \frac{\lambda+1-\mu}{r}\int_0^1 t^{\mu-1}J_{\lambda+1}(rt)\,dt.$$

Further, the first assertion in **2)** immediately follows from applying (4.20) to J_λ and the usual estimate of integral. Now, let $r \geq 1$. After the linear change of variables

$$i(\mu, \lambda, r) = r^{-1-\mu} \int_0^r t^\mu J_\lambda(t)\, dt.$$

Decomposing the integral, we see that it is bounded when $t \in [0,1]$, while for $t \in [1,r]$ we use (4.19). If $\mu > \frac{1}{2}$, we obtain after integrating by parts

$$\begin{aligned}\int_1^r t^\mu J_\lambda(t)\, dt &= \int_1^r t^\mu \left[\sqrt{\frac{2}{\pi}} \frac{\cos(t - \frac{\pi\lambda}{2} - \frac{\pi}{4})}{\sqrt{t}} + O(t^{-\frac{3}{2}})\right] dt \\ &= \sqrt{\frac{2}{\pi}} r^{\mu - \frac{1}{2}} \sin(r - \frac{\pi\lambda}{2} - \frac{\pi}{4}) + O(1) + \int_1^r O(t^{\mu - \frac{3}{2}}) dt \\ &= O\big(r^{\mu - \frac{1}{2}}\big),\end{aligned}$$

and

$$i(\mu, \lambda, r) = O\big(r^{-1-\mu} r^{\mu - \frac{1}{2}}\big) = O\big(r^{-\frac{3}{2}}\big).$$

If $\mu \leq \frac{1}{2}$, then the same computations show that the integral

$$\int_1^r t^\mu J_\lambda(t) dt$$

is bounded with respect to r and

$$i(\mu, \lambda, r) = O\big(r^{-1-\mu}\big).$$

The lemma is proved. □

4.3 Proof of the existence theorem

We are now in a position to prove the above existence Theorem 4.14.

Proof. Let us start with proving (4.15). By Theorem 4.1, for each $\varphi \in \mathcal{S}$,

$$\begin{aligned}\langle \tilde{f}, \varphi \rangle &= \int_{\mathbb{R}^d} \tilde{f}(x)\, \varphi(x)\, dx \\ &= \lim_{A \to \infty} \int_{\mathbb{R}^d} \varphi(x) \left[2\pi |x|^{1-\frac{d}{2}} \int_0^A f_0(t)\, t^{\frac{d}{2}} J_{\frac{d}{2}-1}(2\pi |x| t)\, dt \right] dx.\end{aligned}$$

We rewrite this equality as

$$\langle \tilde{f}, \varphi \rangle = 2\pi \lim_{A \to \infty} \int_0^\infty \varphi_0(r)\, r^{\frac{d}{2}} dr \int_0^A f_0(t)\, t^{\frac{d}{2}} J_{\frac{d}{2}-1}(rt)\, dt.$$

To prove (4.15) we will make use of integration by parts as many times as the right-hand side of (4.15) to be obtained, and then the passage to the limit under the integral sign should be justified. Because of Lemma 4.13 we postpone the fractional differentiation till the last step. Thus, for $[\alpha] \neq 0$ integrating by parts $[\alpha]$ times (in t) yields

$$\int_0^\infty \varphi_0(r) r^{\frac{d}{2}} \int_0^A f_0(t) t^{\frac{d}{2}} J_{\frac{d}{2}-1}(2\pi rt)\, dt\, dr$$
$$= \int_0^\infty \varphi_0(r)\, r^{\frac{d}{2}} \Bigg\{ \frac{(-1)^{[\alpha]}}{([\alpha]-1)!} \int_0^A f_0^{([\alpha])}(t)\, t^{[\alpha]+\frac{d}{2}} dt \int_0^1 (1-s)^{[\alpha]-1} s^{\frac{d}{2}} J_{\frac{d}{2}-1}(2\pi rts)\, ds$$
$$+ \sum_{p=0}^{[\alpha]-1} \frac{(-1)^p}{p!} f_0^{(p)}(t)\, t^{p+\frac{d}{2}+1} \int_0^1 (1-s)^p s^{\frac{d}{2}} J_{\frac{d}{2}-1}(2\pi rts)\, ds \Bigg|_0^A \Bigg\} dr. \tag{4.25}$$

The integrated terms vanish at $t = 0$ by Lemma 4.12 and (4.20).

For $t = A$, we from now on keep in mind that the limit as $A \to \infty$ should be found. Here it suffices, in view of (4.3) and Lemma 4.12, to establish the uniform boundedness in t of the integrals

$$B_p(t) = \int_0^\infty \varphi_0(r)\, r^{\frac{d}{2}} dr t^{\frac{d}{2}+1} \int_0^1 (1-s)^p s^{\frac{d}{2}} J_{\frac{d}{2}-1}(rts)\, ds.$$

For these calculations, we omit 2π in the Bessel function without loss of generality. We integrate by parts in the outer integral $m = [\frac{d}{2} - 1/2]$ times, using (4.18) so that the order of the Bessel function decreases. Integrated terms vanish since $\varphi \in \mathcal{S}$. Defining by ψ here and below the radial part of a function from $\mathcal{S}$, we have

$$B_p(t) = \int_0^\infty \psi(r)\, r^{\frac{d}{2}-m} t^{\frac{d}{2}-m+1} dr \int_0^1 (1-s)^p s^{\frac{d}{2}-m} J_{\frac{d}{2}-m-1}(rts)\, ds.$$

For d odd,

$$B_p(t) = t \int_0^\infty \psi(r)\, dr \int_0^1 (1-s)^p \cos\, rts\, ds.$$

For $p = 0$, we have the Fourier integral formula (see [35, §9]):

$$\lim_{t\to\infty} B_0(t) = \frac{1}{\pi}\psi(0).$$

For $p \geq 1$,

$$\left| t \int_1^\infty \psi(r)\, dr \int_0^1 (1-s)^p \cos\, rts\, ds \right| \leq \int_1^\infty |\psi(r)|\, r^{-1} dr < \infty$$

and

$$\begin{aligned}
t\int_0^1 \psi(r)\, dr \int_0^1 (1-s)^p \cos\, rts\, ds
&= t\int_0^1 \psi(0)\, dr \int_0^1 (1-s)^p \cos\, rts\, ds \\
&+ t\int_0^1 [\psi(r)-\psi(0)]\, dr \int_0^1 (1-s)^p \cos\, rts\, ds.
\end{aligned}$$

The first integral on the right-hand side is equal to

$$\psi(0)\int_0^1 (1-s)^p s^{-1} \sin ts\, ds,$$

and its finiteness is well known. For the second integral, the estimate

$$\int_0^1 |\psi(r)-\psi(0)|\,\frac{dr}{r} < \infty$$

follows, as above. Now, for d even,

$$\begin{aligned}
B_p(t) &= \int_0^\infty \psi(r)\, rt^2 dr \int_0^1 (1-s)^p s J_0(rts)\, ds \\
&= -\int_0^\infty \psi'(r)\left[rt\int_0^1 (1-s)^p s J_1(rts)\, ds\right] dr,
\end{aligned}$$

and integration by parts in accordance with (4.18) yields the boundedness of $B_p(t)$ immediately.

Therefore, the integrated terms in (4.25) vanish as $A\to\infty$. For α integer, the element of integration coincides with that indicated in (4.15).

When α is fractional, apply to the first term in the curly brackets on the right-hand side of (4.25) the following formula of fractional integration by parts (see [11, p. 182], or [167, (2.20)])

$$\int_0^A f_1(t)\, R_\gamma(f_2\,;t)\, dt = \int_0^A W_A^\gamma f_1(t)\, f_2(t)\, dt.$$

In our case $\gamma = 1-\alpha+[\alpha]$, $f_1(t) = f_0^{([\alpha])}(t)$, and

$$f_2(t) = \frac{\Gamma([\alpha])}{\Gamma(\alpha)}\frac{d}{dt}[t^{\alpha+\frac{d}{2}} Q_\alpha(rt)].$$

Indeed,

$$\begin{aligned}
&R_{\alpha-[\alpha]}(s^{[\alpha]+\frac{d}{2}}Q_{[\alpha]}(rs);t)\\
&\quad = \Gamma([\alpha])R_{\alpha-[\alpha]}\left(R_{[\alpha]}(s^{\frac{d}{2}}J_{\frac{d}{2}-1}(rs);\cdot);t\right)\\
&\quad = \Gamma([\alpha])R_{\alpha}(s^{\frac{d}{2}}J_{\frac{d}{2}-1}(rs);t)\\
&\quad = \frac{\Gamma([\alpha])}{\Gamma(\alpha)}t^{\alpha+\frac{d}{2}}Q_{\alpha}(rt).
\end{aligned}$$

Therefore f_2 is the Riemann–Liouville derivative of order $1-\alpha+[\alpha]$ of the function $t^{[\alpha]+\frac{d}{2}}Q_{[\alpha]}(rt)$, and the Riemann–Liouville integral of order $1-\alpha+[\alpha]$ of f_2 is exactly $t^{[\alpha]+\frac{d}{2}}Q_{[\alpha]}(rt)$.

Applying the usual integration by parts to the right-hand side of the formula of the fractional integration by parts gives the following equality true for all α :

$$\begin{aligned}
&\frac{1}{([\alpha]-1)!}\int_0^A f_0^{([\alpha])}(t)\; t^{[\alpha]+\frac{d}{2}}Q_{[\alpha]}(rt)\,dt\\
&= \frac{1}{\Gamma(\alpha)}\int_0^A \frac{d}{dt}[t^{\alpha+\frac{d}{2}}Q_{\alpha}(rt)]\frac{1}{\Gamma(1-\alpha+[\alpha])}\int_t^A f^{([\alpha])}(s)(s-t)^{[\alpha]-\alpha}\,ds\,dt\\
&= \frac{1}{\Gamma(\alpha)\Gamma(1-\alpha+[\alpha])}\left[\int_t^A (s-t)^{-\alpha+[\alpha]}f_0^{([\alpha])}(s)\;ds\right]t^{\alpha+\frac{d}{2}}Q_{\alpha}(rt)\;\Big|_0^A\\
&+\frac{[\alpha]-\alpha}{\Gamma(\alpha)\Gamma(1-\alpha+[\alpha])}\int_0^A t^{\alpha+\frac{d}{2}}Q_{\alpha}(rt)\int_A^\infty (s-t)^{-\alpha+[\alpha]-1}f_0^{([\alpha])}(s)\,ds\,dt\\
&-\frac{1}{\Gamma(\alpha)}\int_0^A F_{\alpha}(t)\;t^{\alpha+\frac{1}{2}}Q_{\alpha}(rt)\;dt.
\end{aligned}$$

Here we obtain the last two values on the right by applying Lemma 4.13. It must be shown again that the first two values on the right-hand side vanish as $A\to\infty$. For t large enough,

$$t^{\frac{d}{2}+\alpha}\left|\int_t^A (s-t)^{-\alpha+[\alpha]}f^{([\alpha])}(s)\,ds\right| \le t^{\frac{d}{2}+\alpha-[\alpha]}\int_t^A (s-t)^{-\alpha+[\alpha]}|s^{[\alpha]}f^{([\alpha])}(s)|\,ds,$$

and because of (4.3) and Lemma 4.12 the right-hand side vanishes at $t=A$. We now consider

$$t^{\frac{d}{2}+\alpha}Q_{\alpha}(rt)\int_t^A (s-t)^{-\alpha+[\alpha]}f^{([\alpha])}(s)\,ds$$

as $t\to 0$. Using the inequality

$$Q_{\alpha}(rt)\le (Cr^{\frac{d}{2}-1})t^{\frac{d}{2}-1}$$

that immediately follows from (4.20), we obtain as above

$$\left| t^{\frac{d}{2}+\alpha} Q_\alpha(rt) \int_1^A (s-t)^{-\alpha+[\alpha]} f^{([\alpha])}(s)\,ds \right|$$
$$\le C(A-t)^{1-\alpha+[\alpha]} t^{d-1+\alpha-[\alpha]} \sup_{[1,A]} \left| s^{[\alpha]} f^{([\alpha])}(s) \right|,$$

and the right-hand side tends to zero as $t \to 0+$. To estimate

$$t^{\frac{d}{2}+\alpha} Q_\alpha(rt) \int_t^1 (s-t)^{-\alpha+[\alpha]} f^{([\alpha])}(s)\,ds,$$

we wish to show that

$$\lim_{t\to 0+} t^{\alpha+d-1} \int_t^1 (s-t)^{[\alpha]-\alpha} f_0^{([\alpha])}(s)\,ds = 0.$$

From the definition of F_α it follows that

$$\frac{d}{dt} \int_t^\infty (s-t)^{[\alpha]-\alpha} f_0^{([\alpha])}(s)\,ds = O\big(t^{\frac{1-d}{2}}\big).$$

Since

$$\frac{d}{dt} \int_1^\infty (s-t)^{[\alpha]-\alpha} f_0^{([\alpha])}(s)\,ds = (\alpha-[\alpha]) \int_1^\infty (s-t)^{[\alpha]-\alpha-1} f_0^{([\alpha])}(s)\,ds$$

is bounded, we have

$$\frac{d}{dt} \int_t^1 (s-t)^{[\alpha]-\alpha} f_0^{([\alpha])}(s)\,ds = O\big(t^{\frac{1-d}{2}}\big)$$

and thus

$$\int_t^1 (s-t)^{[\alpha]-\alpha} f_0^{([\alpha])}(s)\,ds = O\big(t^{\frac{3-d}{2}}\big).$$

This estimate is that desired.

Further, we integrate by parts $[\frac{d-1}{2}]$ times in r. By this, we use (4.18) with "−" on the left-hand side. Using once more (4.18) with "+" on the left-hand side, we obtain

$$\int_0^\infty \varphi_0(r)\, r^{\frac{d}{2}} \int_0^A \left\{ \int_A^\infty (s-t)^{-\alpha+[\alpha]-1} f_0^{([\alpha])}(s)\, ds \right\} t^{\frac{d}{2}+\alpha} Q_\alpha(rt)\, dt\, dr$$

$$= \int_0^\infty \psi(r) r^{\frac{d}{2}-[\frac{d+1}{2}]+1} \int_0^A \left\{ \int_A^\infty (s-t)^{-\alpha+[\alpha]-1} f_0^{([\alpha])}(s)\ ds \right\} t^{\frac{d}{2}-[\frac{d+1}{2}]+\alpha}$$
$$\times \int_0^1 (1-u)^{\alpha-1} u^{\frac{d}{2}-[\frac{d+1}{2}]} J_{\frac{d}{2}+1-[\frac{d+1}{2}]}(rtu)\, du\, dt\, dr.$$

In view of (4.3) and Lemma 4.12, we have

$$\sup_{s\in[A,\infty)} |s^{[\alpha]}\ f_0^{([\alpha])}(s)| \to 0 \quad \text{as} \quad A \to \infty.$$

Besides that,

$$\int_0^A t^{\alpha-[\alpha]-1}\ dt \int_A^\infty (s-t)^{-\alpha+[\alpha]-1}\ ds$$
$$= \frac{1}{\alpha-[\alpha]} \int_0^1 t^{\alpha-[\alpha]-1} (1-t)^{-\alpha+[\alpha]}\ dt \le C.$$

For d odd and each α, since

$$J_{\frac{1}{2}}(rtu) = \sqrt{\frac{2}{\pi rtu}}\ \sin rtu,$$

we derive that

$$\left| \int_0^1 (1-u)^{\alpha-1} u^{\frac{d}{2}-[\frac{d+1}{2}]} J_{\frac{d}{2}+1-[\frac{d+1}{2}]}(rtu)\, dr \right|$$

times $(rt)^{\frac{d}{2}-[\frac{d+1}{2}]+1}$ is bounded. For d even, Lemma 4.17, with $d=2$, yields such an estimate only for $\alpha \ge \frac{1}{2}$. It remains to consider

$$\int_0^\infty \psi(r)\, r \int_0^A \left\{ \int_A^\infty (s-t)^{-\alpha-1} f_0^{([\alpha])}\ (s)\ ds \right\} t^\alpha \int_0^1 (1-u)^{\alpha-1}\ J_1(rtu)\, du\, dt\, dr$$

for $0<\alpha<\frac{1}{2}$. Again, applying Lemma 4.17 with $d=2$, we reduce the problem to the finiteness of

$$\int_1^\infty \psi(r)\ (rt)^{\frac{1}{2}-\alpha} \sin rt\ dr$$

when $rt>1$, since all the remainder terms are estimated as above by applying (4.18). Integrating by parts and taking into account that all the integrated terms are bounded, we see that it remains to estimate

$$\int_1^\infty \psi'(r)\ r(rt)^{-\frac{1}{2}-\alpha} \cos rt\, dr + \left(\frac{1}{2}-\alpha\right) \int_1^\infty \psi(r)\ (rt)^{-\frac{1}{2}-\alpha} \cos rt\, dr.$$

Since both integrals are finite, we finally get

$$\langle \tilde{f}, \varphi \rangle = \frac{2\pi(-1)^{\alpha^*+1}}{\Gamma(\alpha)} \lim_{A\to\infty} \int_0^\infty \varphi_0(r) r^{\frac{d}{2}} \int_0^A F_\alpha(t) t^{\alpha+\frac{1}{2}} Q_\alpha(rt)\, dt\, dr.$$

It remains to justify the passage to the limit under the integral sign. Due to the Lebesgue dominated convergence theorem, it is possible if the element of integration is dominated by an integrable function independent of A. To prove this, we consider two integrals: over $r \in [0,1]$ and over $r \in [1,\infty)$, respectively. In view of (4.3) and (4.4), one can treat F_α as a function which is monotone decreasing and vanishing at infinity. Let $r \in [1,\infty)$. Consider two integrals in t over$[0,1]$ and $[1,A]$, respectively. The first one is bounded, and the bound is simply $|\varphi_0(r)\ r^{\frac{d}{2}}|$ times absolute constant. Applying the second mean value theorem of integral calculus to the integral over $[1,A]$ and using (4.18) and Lemma 4.17, we obtain ($\xi \le A$)

$$\begin{aligned}
&\left|\varphi_0(r)\ r^{\frac{d}{2}} \int_1^A F_\alpha(t)\ t^{\alpha+\frac{1}{2}} Q_\alpha(rt)\ dt\right| \\
&\quad = \left|\varphi_0(r)\ r^{\frac{d}{2}} F_\alpha(1) \int_1^\xi t^{\alpha+\frac{1}{2}} Q_\alpha(rt)\ dt\right| \\
&\quad = \left|F_\alpha(1)\ \varphi_0(r)\ r^{\frac{d}{2}-1} \left\{ t^{\alpha+1/2}\ q_\alpha(rt)\ \Big|_1^\xi + \left(\frac{d-1}{2} - \alpha\right) \int_1^\xi t^{\alpha-\frac{1}{2}}\ q_\alpha(rt)\ dt \right\}\right| \\
&\quad \le \left(\frac{d-1}{2} - \alpha\right) \left|F_\alpha(1)\ \varphi_0(r)\ r^{\frac{d-3}{2}-\alpha} \int_1^\xi \frac{1}{t} \cos(rt+\mu)\, dt\right| + C\frac{1}{r}\ |\ F_\alpha(1)\ \varphi_0(r)\ |,
\end{aligned}$$

and the last value is integrable over $[1,\infty)$. Finally,

$$\begin{aligned}
&\lim_{A\to\infty} \int_0^1 \varphi_0(r)\ r^{\frac{d}{2}} dr \int_0^A F_\alpha(t)\ t^{\alpha+\frac{1}{2}} Q_\alpha(rt)\, dt \\
&\quad = \lim_{A\to\infty} \int_0^1 \{[\varphi_0(r) - \varphi_0(0)] + \varphi_0(0)\}\, r^{\frac{d}{2}}\ dt \int_0^A F_\alpha(t)\ t^{\alpha+\frac{1}{2}} Q_\alpha(rt)\ dt.
\end{aligned}$$

Since $\frac{1}{r}[\varphi_0(r) - \varphi_0(0)]$ is integrable, the part corresponding to this function may be estimated completely like that in the case $r \in [1,\infty)$. Further,

$$\lim_{A\to\infty} \int_0^1 r^{\frac{d}{2}} \int_0^A F_\alpha(t)\ t^{\alpha+\frac{1}{2}} Q_\alpha(rt)\, dt\, dr = \lim_{A\to\infty} \int_0^A F_\alpha(t)\ t^{\alpha-\frac{1}{2}}\ q_a(t)\, dt.$$

Since F_α is bounded and monotone, integration by parts and estimates like in Lemma 4.17 yield the convergence of this integral in improper sense.

In fact, we have proved the uniform convergence of the integral (4.15) when $|x| \ge r_0 > 0$.

Let us now show that $\widehat{f}(x) \to 0$ as $r = |x| \to \infty$. Using the second mean value theorem, we obtain for some $A'' \le A'$,

$$r^{1-\frac{d}{2}} \int_A^{A'} F_\alpha(t)\ t^{\alpha+\frac{1}{2}}\ Q_\alpha(rt)\, dt = F_\alpha(A)\ r^{-\frac{d}{2}} \left[t^{\alpha+\frac{1}{2}}\ q_\alpha(rt)\right]_A^{A''}.$$

In view of Lemma 4.17, we have for every $A \ge 1$ as $A' \to \infty$,

$$\left| r^{1-\frac{d}{2}} \int_A^\infty F_\alpha(t)\, t^{\alpha+\frac{1}{2}}\, Q_\alpha(rt)\, dt \right| \le C \mid F_\alpha(A) \mid r^{-\frac{d-1}{2}-\alpha}.$$

The right-hand side tends to zero as $r \to \infty$. Further,

$$\begin{aligned} &r^{1-\frac{d}{2}} \int_0^A F_\alpha(t)\; t^{\alpha+\frac{1}{2}}\; Q_\alpha(rt)\; dt \\ &\quad = F_\alpha(t)\; r^{-\frac{d}{2}}\; q_\alpha(rt)\; \Big|_0^A - r^{-\frac{d}{2}} \int_0^A t^{\alpha+\frac{1}{2}}\; q_\alpha(rt)\; dF_\alpha(t) \\ &\quad + \left(\frac{d-1}{2} - \alpha\right) r^{-\frac{d}{2}} \int_0^A F_\alpha(t)\; t^{\alpha-\frac{1}{2}}\; q_\alpha(rt)\; dt = O\big(r^{-\frac{d}{2}}\big) \end{aligned}$$

by Lemma 4.17 and (4.4), which completes the proof.

Let us show the continuity of $\widehat{f}(x)$ for $|x| > 0$. Let $[r_0, r_1]$ be an interval of uniform convergence of the integral in (4.15), and $|x| \in [r_0, r_1]$. Then the functions

$$\widehat{f_k}(x) = \frac{(2\pi)^{\frac{d}{2}}(-1)^{\alpha^*+1}}{\Gamma(\alpha)}\; |x|^{1-\frac{d}{2}} \int_0^k F_\alpha(t)\, t^{\alpha+\frac{1}{2}}\, Q_\alpha(|x|t)\, dt$$

are continuous for each $k = 1, 2, \ldots,$ and converge uniformly to $\widehat{f}(x)$ as $k \to \infty$. Hence, $\widehat{f}(x)$ is continuous for these x as well.

Let us now prove the inverse formula. Applying the Cauchy–Poisson formula (see Theorem 1.6 above), we have

$$\begin{aligned} &\frac{1}{(2\pi)^2} \int_{|u|\le A} \left(1 - \frac{|u|^2}{A^2}\right)^{\frac{d-1}{2}-\alpha} \widehat{f}(u)\; e^{2\pi i x\cdot u}\; du \\ &= \frac{(-1)^{\alpha^*+1}}{\Gamma(\alpha)}\; r^{1-\frac{d}{2}} \int_0^A \left(1 - \frac{s^2}{A^2}\right)^{\frac{d-1}{2}-\alpha} s J_{\frac{d}{2}-1}(2\pi r s)\; ds \\ &\times \int_0^\infty F_\alpha(t)\; t^{\alpha+\frac{1}{2}}\; Q_\alpha(2\pi s t)\; dt \\ &= \frac{(-1)^{\alpha^*+1}}{\Gamma(\alpha)}\; r^{1-\frac{d}{2}} \int_0^\infty F_\alpha(t)\; t^{\alpha+\frac{1}{2}}\; dt \\ &\times \int_0^A \left(1 - \frac{s^2}{A^2}\right)^{\frac{d-1}{2}-\alpha} s J_{\frac{d}{2}-1}(2\pi r s)\; Q_\alpha(2\pi s t)\; ds. \end{aligned} \tag{4.26}$$

Here, changing the order of integration must be justified. Omitting 2π when needed in the consequent calculations results in no loss of generality. Let $0 < \delta < A$. The uniform convergence of the integral in t for $s \geq \delta$ yields

$$\int_\delta^A \left(1-\frac{s^2}{A^2}\right)^{\frac{d-1}{2}-\alpha} sJ_{\frac{d}{2}-1}(rs)\,ds \int_0^\infty F_\alpha(t)\, t^{\alpha+\frac12}\, Q_\alpha(st)\,dt$$
$$= \int_0^\infty F_\alpha(t)\, t^{\alpha+\frac12}\,dt \int_0^A \left(1-\frac{s^2}{A^2}\right)^{\frac{d-1}{2}-\alpha} sJ_{\frac{d}{2}-1}(rs)\, Q_\alpha(st)\,ds$$
$$- \int_0^\infty F_\alpha(t)\, t^{\alpha+\frac12}\,dt \int_0^\delta \left(1-\frac{s^2}{A^2}\right)^{\frac{d-1}{2}-\alpha} sJ_{\frac{d}{2}-1}(rs)\, Q_\alpha(st)\,ds. \qquad (4.27)$$

It suffices to show that the last integral tends to zero as $\delta \to 0$. Take $\varepsilon > 0$ and let M be large enough to provide $|F_\alpha(M)| < \varepsilon$, by (4.3). The second mean value theorem yields, after integrating by parts,

$$\int_M^\infty F_\alpha(t)\, t^{\alpha+\frac12} \int_0^\delta \left(1-\frac{s^2}{A^2}\right)^{\frac{d-1}{2}-\alpha} sJ_{\frac{d}{2}-1}(rs)\, Q_\alpha(st)\,ds\,dt$$
$$= F_\alpha(M) \int_M^{M'} t^{\alpha+\frac12} \int_0^\delta \left(1-\frac{s^2}{A^2}\right)^{\frac{d-1}{2}-\alpha} sJ_{\frac{d}{2}-1}(rs)\, Q_\alpha(st)\,ds\,dt$$
$$= F_\alpha(M) \left[t^{\alpha+\frac12} \int_0^\delta \left(1-\frac{s^2}{A^2}\right)^{\frac{d-1}{2}-\alpha} J_{\frac{d}{2}-1}(rs)\, q_\alpha(st)\,ds\right]_M^{M'} \qquad (4.28)$$
$$- F_\alpha(M)\left(\alpha-\frac{d-1}{2}\right) \int_M^{M'} t^{\alpha-\frac12} \int_0^\delta \left(1-\frac{s^2}{A^2}\right)^{\frac{d-1}{2}-\alpha} J_{\frac{d}{2}-1}(rs)\, q_\alpha(st)\,ds\,dt.$$

We first estimate the integrated terms in (4.28). The uniform boundedness, in t and δ, of the value in brackets should be shown. Since

$$\zeta_{\alpha,d} \int_0^\delta \left(1-\frac{s^2}{A^2}\right)^{\frac{d-1}{2}-\alpha} s^{-\alpha-\frac12}\, J_{\frac{d}{2}-1}(rs)\,ds$$

does not depend on M and M', these values taken twice with opposite signs cancel one another. In view of (4.20), the rest, for $s \in [0, \frac1t]$, does not exceed the quantity

$$Ct^{\alpha+\frac12} \int_0^{\frac1t} |J_{\frac{d}{2}-1}(rs)|\,ds \leq Cr^{\frac{d}{2}-1}\, t^{\alpha-\frac{d-1}{2}},$$

while for $t\delta > 1$ and $s \in [\frac1t, \delta]$, in view of Lemma 4.17 and (4.20), it is

$$\Gamma(\alpha) t^{\frac12} \int_{\frac1t}^\delta s^{-\alpha} \left(1-\frac{s^2}{A^2}\right)^{\frac{d-1}{2}-\alpha} J_{\frac{d}{2}-1}(rs) J_{\frac{d}{2}+\alpha}(st)\,ds + O\left(t^{-1}\int_{\frac1t}^\delta s^{\frac{d-5}{2}-\alpha} ds\right).$$

The remainder term does not exceed

$$C\{t^{\alpha-\frac{d-1}{2}}+(t\delta)^{-1}\delta^{\frac{d-1}{2}-\alpha}\},$$

which is bounded. Applying (4.20) to $J_{\frac{d}{2}-1}$ and (4.19) to $J_{\frac{d}{2}+\alpha}$ in the leading term, we obtain for $\alpha<\frac{d-1}{2}$,

$$\int_{\frac{1}{t}}^{\delta}\left(1-\frac{s^2}{A^2}\right)^{\frac{d-1}{2}-\alpha}s^{-\alpha+\frac{d-3}{2}}\,ds\le C(\delta^{\frac{d-1}{2}-\alpha}+t^{\alpha-\frac{d-1}{2}})\le C.$$

For $\alpha=\frac{d-1}{2}$, we first integrate by parts as follows:

$$\begin{aligned}&t^{\frac{1}{2}}\int_{\frac{1}{t}}^{\delta}s^{-\frac{d-1}{2}}\,J_{\frac{d}{2}-1}(rs)J_{d-\frac{1}{2}}(st)\,ds\\&=-t^{-\frac{1}{2}}s^{\frac{1-d}{2}}J_{\frac{d}{2}-1}(rs)\,J_{d+\frac{1}{2}}(st)\Big|_{\frac{1}{t}}^{\delta}+rt^{-\frac{1}{2}}\int_{\frac{1}{t}}^{\delta}s^{-\frac{d-1}{2}}J_{\frac{d}{2}-2}\,(rs)\,J_{d+\frac{1}{2}}\,(st)\,ds,\end{aligned}$$

and then continue the proof as that for the remainder term.

Let us now estimate the last integral in (4.28). This makes sense only for $\alpha<\frac{d-1}{2}$. Using again Lemma 4.17, we arrive to estimating

$$\int_{M}^{M'}t^{-\frac{1}{2}}dt\int_{\frac{1}{t}}^{\delta}\left(1-\frac{s^2}{A^2}\right)^{\frac{d-1}{2}-\alpha}s^{-\alpha}J_{\frac{d}{2}-1}\,(rs)\,J_{\frac{d}{2}+\alpha}\,(ts)\,ds,$$

since the rest is treated analogously. Let us change the order of integration. Without loss of generality, one can take $\delta<\frac{1}{M}$. The following should be estimated:

$$\begin{aligned}&\int_{\frac{1}{M}'}^{\delta}\left(1-\frac{s^2}{A^2}\right)^{\frac{d-1}{2}-\alpha}s^{-\alpha}J_{\frac{d}{2}-1}\,(rs)\,ds\int_{\frac{1}{s}}^{M'}t^{-\frac{1}{2}}J_{\frac{d}{2}+\alpha}(st)\,dt\\&+\int_{\delta}^{\frac{1}{M}}\left(1-\frac{s^2}{A^2}\right)^{\frac{d-1}{2}-\alpha}s^{-\alpha}J_{\frac{d}{2}-1}(rs)\,ds\int_{M}^{\frac{1}{s}}t^{-\frac{1}{2}}J_{\frac{d}{2}+\alpha}\,(st)\,dt.\end{aligned}$$

We apply (4.19) to $J_{\frac{d}{2}+\alpha}$. For the remainder term, after applying (4.20) to $J_{\frac{d}{2}-1}$, we obtain

$$\int_{\frac{1}{M}'}^{\delta}s^{\frac{d-5}{2}-\alpha}\int_{\frac{1}{s}}^{M'}\frac{1}{t^2}\,dt\,ds+\int_{\delta}^{\frac{1}{M}}s^{\frac{d-5}{2}-\alpha}\int_{M}^{\frac{1}{s}}\frac{1}{t^2}dt\,ds,$$

which is obviously bounded. For the leading term, the integral in t is of the form

$$\int\frac{\cos(st+\mu)}{t}\,dt.$$

After integrating by parts the estimates coincide with those for the remainder term. Hence, the value (4.28) is small. Choosing δ so small that

$$\left| \int_0^M F_\alpha(t)\, t^{\alpha+\frac{1}{2}} \int_0^\delta \left(1-\frac{s^2}{A^2}\right)^{\frac{d-1}{2}-\alpha} sJ_{\frac{d}{2}-1}(rs)\, Q_\alpha(st)\, ds\, dt \right| < \varepsilon,$$

we derive that the last integral in (4.27) tends to zero as $\delta \to 0$. Returning to (4.26), we have

$$\begin{aligned}
&\int_0^A \left(1-\frac{s^2}{A^2}\right)^{\frac{d-1}{2}-\alpha} sJ_{\frac{d}{2}-1}(2\pi rs) Q_\alpha(2\pi st)\, ds \\
&= \int_0^A sJ_{\frac{d}{2}-1}(rs) Q_\alpha(st)\, ds \\
&\quad - \int_0^A sJ_{\frac{d}{2}-1}(2\pi rs) Q_\alpha(2\pi st)\, 2\left(\frac{d-1}{2}-\alpha\right) \int_0^{\frac{s}{A}} u(1-u^2)^{\frac{d-3}{2}-\alpha} du\, ds.
\end{aligned}$$

Let us proceed to the second integral. We have

$$\begin{aligned}
&\int_0^\infty F_\alpha(t)\, t^{\alpha+\frac{1}{2}} \left[\int_0^A sJ_{\frac{d}{2}-1}(rs) Q_\alpha(st) \int_0^{s/A} u(1-u^2)^{\frac{d-3}{2}-\alpha} du\, ds\right] dt \\
&= \int_0^1 u(1-u^2)^{\frac{d-3}{2}-\alpha} \int_0^\infty F_\alpha(t) t^{\alpha+\frac{1}{2}} \int_{Au}^A sJ_{\frac{d}{2}-1}(rs) Q_\alpha(st)\, ds\, dt\, du \\
&= \int_0^1 u(1-u^2)^{\frac{d-3}{2}-\alpha} \Bigg\{ F_\alpha(t)\, t^{\alpha+\frac{1}{2}} \int_{Au}^A J_{\frac{d}{2}-1}(rs)\, q_\alpha(st)\, ds \Bigg|_0^\infty \\
&\quad - \int_0^\infty t^{\alpha+\frac{1}{2}} \int_{Au}^A J_{\frac{d}{2}-1}(rs)\, q_\alpha(st)\, ds\, dF_\alpha(t) \\
&\quad + \left(\frac{d-1}{2}-\alpha\right) \int_0^\infty F_\alpha(t)\, t^{\alpha-\frac{1}{2}} \int_{Au}^A J_{\frac{d}{2}-1}(rs) q_\alpha(st)\, ds\, dt \Bigg\} du.
\end{aligned}$$

The integrated terms vanish at $t=0$ and $t=\infty$. Indeed, it is obvious for $t=0$, while for $t=\infty$ follows from Lemma 4.17 and (4.3). We have to show that the right-hand side tends to zero as $A \to \infty$. Observe first that the estimate

$$|q_\alpha(st)| \le C(st)^{-\alpha-\frac{1}{2}}$$

as well as (4.19) and (4.20) for $J_{\frac{d}{2}-1}$ yield

$$\left| t^{\alpha+\frac{1}{2}} \int_{Au}^A J_{\frac{d}{2}-1}(rs)\, q_\alpha(st)\, ds \right| \le C \int_{Au}^A s^{-1-\varepsilon} ds$$

with some $\varepsilon \in (0,1)$. This combined with (4.4) gives proper estimates for the second summand in the curly brackets. When $t \in [0,1]$ the calculations for the third one are similar. Using then Lemma 4.17 when $t \in [1,\infty)$, we estimate the remainder term as above. The leading term, by the second mean value theorem, is equal to

$$\int_1^\infty F_\alpha(t)\, t^{-\frac{1}{2}} \int_{Au}^A s^{-\alpha} J_{\frac{d}{2}-1}(rs) J_{\frac{d}{2}+\alpha}(st)\, ds\, dt$$
$$= F_\alpha(1) \int_1^\xi t^{-\frac{1}{2}} \int_{Au}^A s^{-\alpha} J_{\frac{d}{2}-1}(rs) J_{\frac{d}{2}+\alpha}(st)\, ds\, dt.$$

We apply (4.19) to $J_{\frac{d}{2}+\alpha}$. No new technique is needed for the remainder term. What should now be estimated is

$$\int_1^\xi t^{-1} \int_{Au}^A s^{-\alpha-\frac{1}{2}} J_{\frac{d}{2}-1}(rs) \cos(ts+\mu)\, ds\, dt.$$

Integration by parts in t and estimates of the integral in s, like above, prove that the limit is zero. It remains to consider

$$\int_0^\infty F_\alpha(t)\, t^{\alpha+\frac{1}{2}} \int_0^A s J_{\frac{d}{2}-1}(2\pi rs)\; Q_\alpha(2\pi st)\, ds\, dt$$
$$= \frac{1}{2\pi} \int_0^\infty F_\alpha(t)\, t^{\alpha-\frac{d-1}{2}} \frac{d}{dt}\left[t^{\frac{d}{2}} \int_0^1 (1-u)^{\alpha-1} u^{\frac{d}{2}-1} du \int_0^A J_{\frac{d}{2}-1}(2\pi rs) J_{\frac{d}{2}}(2\pi uts)\, ds \right] dt.$$

Let us substitute integration over $[0, A]$ for that over the difference of two sets: $[0,\infty)$ and $[A,\infty)$, and use the formula

$$\Gamma(\nu-\mu) \int_0^\infty J_\mu(at) J_\nu(bt) t^{\mu-\nu+1} dt = \begin{cases} 2^{\mu-\nu+1} a^\mu b^{-\nu} (b^2-a^2)^{\nu-\mu+1}, & \text{for } b > a, \\ 0, & \text{for } b < a, \end{cases}$$

which is true for $\nu > \mu > -1$ (see [12, p. 148]). We obtain

$$\frac{1}{2\pi} \int_0^\infty F_\alpha(t)\, t^{\alpha-\frac{d-1}{2}} \frac{d}{dt}\left[t^{\frac{d}{2}} \int_0^1 (1-u)^{\alpha-1} u^{\frac{d}{2}-1} \int_0^\infty J_{\frac{d}{2}-1}(2\pi rs) J_{\frac{d}{2}}(2\pi uts)\, ds\, du \right] dt$$
$$= \int_r^\infty F_\alpha(t)\, t^{\alpha-\frac{d-1}{2}} \frac{d}{dt}\left[t^{\frac{d}{2}} \int_{r/t}^1 (1-u)^{\alpha-1} u^{\frac{d}{2}-1} r^{\frac{d}{2}-1} (tu)^{-\frac{d}{2}} du \right] dt$$
$$= r^{\frac{d}{2}-1} \int_r^\infty f_0^{(\alpha)}(t)\; (t-r)^{\alpha-1} dt.$$

Integrating by parts and using (4.2) and (4.3), we get

$$\frac{(-1)^{\alpha^*+1}}{\Gamma(a)} \int_r^\infty f_0^{(\alpha)}(t)\; (t-r)^{\alpha-1} dt = f_0(r).$$

For α fractional we have also used the permutability of the fractional integral and fractional derivative. If we show that

$$\int_0^\infty F_\alpha(t)\, t^{\alpha-\frac{d-1}{2}} \frac{d}{dt}\left[t^{\frac{d}{2}} \int_0^1 (1-u)^{\alpha-1} u^{\frac{d}{2}-1} \int_A^\infty J_{\frac{d}{2}-1}(rs) J_{\frac{d}{2}}(uts)\, ds\, du\right] dt$$

tends to zero as $A \to \infty$, the inverse formula will be proved. Integration by parts in the outer integral yields

$$\begin{aligned} &F_\alpha(t)\, t^{\alpha+\frac{1}{2}} \int_0^1 (1-u)^{\alpha-1} u^{\frac{d}{2}-1} \int_A^\infty J_{\frac{d}{2}-1}(rs) J_{\frac{d}{2}}(ust)\, ds\, du \Big|_0^\infty \\ &- \int_0^\infty \left[t^{\alpha+\frac{1}{2}} \int_0^1 (1-u)^{\alpha-1} u^{\frac{d}{2}-1} \int_A^\infty J_{\frac{d}{2}-1}(rs) J_{\frac{d}{2}}(ust)\, ds\, du\right] dF_\alpha(t) \\ &+ \left(\frac{d-1}{2} - \alpha\right) \int_0^\infty F_\alpha(t)\, t^{\alpha-\frac{1}{2}} \\ &\quad \times \int_0^1 (1-u)^{\alpha-1} u^{\frac{d}{2}-1} \int_A^\infty J_{\frac{d}{2}-1}(rs) J_{\frac{d}{2}}(ust)\, ds\, du\, dt. \end{aligned}$$

To estimate the first two summands, it suffices, in view of (4.4), to show that

$$\lim_{A\to\infty} \sup_t \left| t^{\alpha+1/2} \int_0^1 (1-u)^{\alpha-1} u^{\frac{d}{2}-1} du \int_A^\infty J_{\frac{d}{2}-1}(rs) J_{\frac{d}{2}}(ust)\, ds \right| = 0.$$

Let us show that the change of the order of integration is legal. By the Lebesgue dominated convergence theorem, it suffices to find an integrable majorant, independent of A and A', for the function

$$(1-u)^{\alpha-1} u^{\frac{d}{2}-1} \int_A^{A'} J_{\frac{d}{2}-1}(rs) J_{\frac{d}{2}}(uts)\, ds.$$

Integrating by parts in the inner integral, we obtain

$$\int_A^{A'} J_{\frac{d}{2}-1}(rs)\, J_{\frac{d}{2}}(uts)\, ds = \frac{1}{r} J_{\frac{d}{2}}(rs)\, J_{\frac{d}{2}}(uts) \Big|_A^{A'} + \frac{tu}{r} \int_A^{A'} J_{\frac{d}{2}}(rs)\, J_{\frac{d}{2}+1}(uts)\, ds.$$

The integrated terms are bounded, and $(1-u)^{\alpha-1} u^{\frac{d}{2}-1}$ is integrable over $[0,1]$. We now apply (4.19) to the Bessel functions in the last integral. The remainder terms are bounded, and $(1-u)^{\alpha-1} u^{\frac{d-3}{2}}$ is integrable. For the leading terms, we have

$$\begin{aligned} &\int_A^{A'} \frac{\cos(rs - \frac{\pi d}{2} - \frac{\pi}{4}) \cos(uts - \frac{\pi d}{2} - \frac{3\pi}{4})}{s}\, ds \\ &\quad = \frac{1}{2} \int_A^{A'} \frac{\cos(rs + uts - \pi d - \pi)}{s}\, ds - \frac{1}{2} \int_A^{A'} \frac{\sin s(r - tu)}{s}\, ds, \end{aligned}$$

and the boundedness of these integrals is easily obtained by integration by parts and by the inequality $\frac{1}{r+ut} \le \frac{1}{r}$. Now, Lemma 4.17 and the rough estimate $J_\nu(t) = O(t^{-\frac{1}{2}})$ yield

$$\lim_{A\to\infty} \sup_t \left| t^{\alpha+\frac{1}{2}} \int_0^1 (1-u)^{\alpha-1} u^{\frac{d}{2}-1} du \int_A^\infty J_{\frac{d}{2}-1}(rs)\, J_{\frac{d}{2}}(uts)\, ds \right|$$
$$= \lim_{A\to\infty} \sup_t \left| t^{\alpha+\frac{1}{2}} \int_A^\infty J_{\frac{d}{2}-1}(rs)\, q_\alpha(ts)\, ds \right| \le C \lim_{A\to\infty} r^{-\frac{1}{2}} \int_A^\infty \frac{ds}{s^{1+\alpha}} = 0.$$

It remains to estimate

$$\int_0^\infty F_\alpha(t)\, t^{\alpha-\frac{1}{2}} dt \int_0^1 (1-u)^{\alpha-1} u^{\frac{d}{2}-1} du \int_A^\infty J_{\frac{d}{2}-1}(rs)\, J_{\frac{d}{2}}(uts)\, ds$$
$$= \int_0^\infty F_\alpha(t)\, t^{\alpha-\frac{1}{2}} dt \int_A^\infty J_{\frac{d}{2}-1}(rs)\, q_\alpha(ts)\, ds.$$

For $t \in [\frac{r}{2}, \infty]$ such estimates are already fulfilled above. Let $t \in [0, \frac{r}{2}]$. Considering

$$\int_0^{\frac{r}{2}} F_\alpha(t) t^{\alpha-\frac{1}{2}} \int_0^1 (1-u)^{\alpha-1} u^{\frac{d}{2}-1} \int_A^\infty J_{\frac{d}{2}-1}(rs) J_{\frac{d}{2}}(ust)\, ds\, du\, dt,$$

we apply (4.19) to the first Bessel function on the right-hand side. Estimates for the remainder terms are obvious, therefore we have to estimate

$$\int_0^{\frac{r}{2}} F_\alpha(t) t^{\alpha-\frac{1}{2}} \int_0^1 (1-u)^{\alpha-1} u^{\frac{d}{2}-1} \int_A^\infty s^{-\frac{1}{2}} \cos(rs-h) J_{\frac{d}{2}}(ust)\, ds\, du\, dt$$

where h is a number. Integration by parts in the integral

$$\int_A^\infty s^{-\frac{1}{2}} \cos(rs-h) J_{\frac{d}{2}}(ust)\, ds = \int_A^\infty s^{-\frac{d}{2}} s^{-\frac{1}{2}} \cos(rs-h)\, s^{\frac{d}{2}} J_{\frac{d}{2}}(ust)\, ds$$

and simple calculations using (4.18) and (4.19) lead to showing that the integral

$$ut \int_A^\infty J_{\frac{d}{2}-1}(ust) s^{-\frac{1}{2}} \sin(rs-h)\, ds$$

tends to zero as $A \to \infty$. We are going to apply (4.19) to the Bessel function. This is the reason that the factor ut precedes the integral. It ensures that the other integrals, in t and u, exist. Again, for the remainder terms, we get estimates similar to those above by applying (4.19). For the main term, we have the inner integral of the form

$$\sqrt{ut}\int\limits_A^\infty \frac{\cos(ust-l)\sin(rs-h)}{s}\,ds.$$

Since $ut \le \frac{r}{2}$ for $t \in [0, \frac{r}{2}]$ and $u \in [0,1]$, the latter integral is obviously small as $A \to \infty$. The proof is complete. □

4.4 Passage to a one-dimensional Fourier transform

What we have achieved up to now is the possibility to represent the Fourier transform of a radial function from an important space as a regular function, more precisely, as a convergent improper Hankel type integral, though initially it should be treated only in the distributional sense. Of course, importance of the spaces in question needs additional discussion, but this will be postponed to the consideration of applications. What is vague now is that the convenience of those Hankel type integrals is disputable.

It turns out that we are now in a position to carry out the passage from the multidimensional Fourier transform to the one-dimensional Fourier transform of a related function. The result reads as follows.

Theorem 4.29. *Let $f_0 \in MV^b_{\alpha+1}$ with $0<\alpha\le\frac{d-1}{2}$ and $b=\frac{d-1}{2}-\alpha$, set*

$$F_\alpha(t) = t^{\frac{d-1}{2}} f_0^{(\alpha)}(t).$$

Then there holds, for the radial extension $f(x)=f_0(|x|)$ of f_0, with $|x|>0$,

$$\begin{aligned} f(x) = |x|^{-\frac{d-1}{2}-\alpha}\Big\{ C_{1,d}\int_0^\infty F_\alpha(t)\cos\Big(2\pi|x|t-\frac{\pi}{4}(d+2\alpha-1)\Big)\,dt \\ + C_{2,d}\frac{1}{|x|}F_\alpha\Big(\frac{1}{|x|}\Big) + O\Big(\frac{1}{|x|}\int_0^\infty \min\Big(|x|s,\frac{1}{|x|s}\Big)|dF_\alpha(s)|\Big)\Big\}, \end{aligned} \tag{4.30}$$

where $C_{1,d} = 2^{\frac{3}{2}}\pi^{\frac{1}{2}}(-1)^{\alpha^*+1}$ and $C_{2,d}$ is a constant depending only on d and α, given in (4.34) *below.*

Remark 4.31. Concerning the terms in the brackets on the right-hand side of (4.15), the first one is the claimed one-dimensional Fourier transform, unlike more complicated integrals in (1.6) or (4.15). In a concrete situation, an explicit formula may exist – there are numerous detailed tables (see, *e.g.*, [12]). The integral exists as an improper Riemann integral, since $F_\alpha \in BV[0,\infty)$ and vanishes at infinity; this is a direct generalization of a well-known one-dimensional result (see, *e.g.*, [31, § 2]).

The second term in the brackets on the right side of (4.30), being given explicitly, is easy to handle. Its L^p-integrability reduces to

$$\int_1^N r^{n-1}\left|r^{-\frac{d-1}{2}-\alpha-1}F_\alpha\left(\frac{1}{r}\right)\right|^p dr = C\int_{\frac{1}{N}}^1 r^{-d-1+p\frac{d}{2}-\frac{p}{2}+p\alpha+p}|F_\alpha(r)|^p\,dr.$$

The integral is uniformly bounded, with respect to N, if $\alpha > \frac{d}{p}-\frac{d-1}{2}$. Thus the second term on the right-hand side of (4.30) is essential, that is, may lead to a non-L^p-integrable term, if $\alpha \le \frac{d}{p}-\frac{d-1}{2}$.

If we choose, in particular, $p=1$ and $\alpha=\frac{d-1}{2}$, we end up with the integral

$$\int_{\frac{1}{N}}^1 |F(t)|\,\frac{dt}{t}, \qquad F(t)=t^{\frac{d-1}{2}}f_0^{(\frac{d-1}{2})}(t), \tag{4.32}$$

which may diverge under the hypothesis $f_0\in MV_{\frac{d+1}{2}}$ as examples, more or less similar to those in Proposition 4.9, show. Let, for instance, $d=3$. Consider the function

$$f(x)=\sin\left(\ln\ln\left(\frac{e}{|x|}\right)\right)$$

for $|x|\in[0,1]$, and 0 otherwise. We have

$$f_0'(t)=\frac{1}{t\ln\left(\frac{e}{t}\right)\ln\ln\left(\frac{e}{t}\right)}\cos\left(\ln\ln\left(\frac{e}{t}\right)\right)$$

and $F(t)=tf_0'(t)$. This function obviously satisfies conditions (4.2)–(4.4). It is easy to see that

$$\int_0^1\frac{|F(t)|}{t}\,dt=\int_0^1|f'(t)|\,dt=\infty.$$

Consider

$$\begin{aligned}\int\limits_{1\le|x|\le N}|\widehat{f}(x)|\,dx &= C\int_1^N r^{\frac{3}{2}}\left|\int_0^1 F(t)t^{\frac{3}{2}}\int_0^1 s^{\frac{3}{2}}J_{\frac{1}{2}}(rts)\,ds\,dt\right|dr\\ &= C\sqrt{\frac{2}{\pi}}\int_1^N r\left|\int_0^1 F(t)t\int_0^1 s\sin rts\,ds\,dt\right|dr\\ &= C\sqrt{\frac{2}{\pi}}\int_1^N\left|\int_0^1 F(t)\cos rt\,dt-\int_0^1 F(t)\frac{\sin rt}{rt}\,dt\right|dr.\end{aligned}$$

It suffices now to prove that

$$\lim_{N\to\infty}\int_1^N\frac{1}{r}\left|\int_0^1 F(t)\frac{\sin rt}{t}\,dt\right|dr=\infty.$$

Integrating by parts, we have

$$\int_1^N \left| \int_0^1 t^{-1} F(t) \sin rt\, dt \right| \frac{dr}{r} = \int_1^N \left| \int_0^1 f_0(t) \cos rt\, dt \right| dr,$$

and one has to prove that the one-dimensional Fourier transform of the function f_0 is non-integrable. Indeed, if it were integrable, the following condition would necessarily be valid (see, *e.g.*, [113, Ch. 2, §10]): the integral

$$\int_0^1 \frac{f_0(t)}{t}\, dt$$

converges. But it is easy to see that this integral diverges for our function, which shows the sharpness of the condition (4.32).

The integral in the last term of the right-hand side of (4.30) is in any case dominated by $\|f_0\|_{MV^b_{\alpha+1}}$. Concerning the L^p-integrability, it turns out that in the case $p = 1$ and $\alpha = \frac{d-1}{2}$ it is finite (this is the case considered in [120]; see also [22, 23]) whereas in the case $\alpha \le \frac{d}{p} - \frac{d-1}{2}$ it may be as bad/good as the second term. To see this, we apply the generalized Minkowski inequality and obtain

$$\left\{ \int_1^N r^{d-1} \left| r^{-\frac{d-1}{2}-\alpha-1} \int_0^\infty \min\left(rs, \frac{1}{rs}\right) |dF_\alpha(s)| \right|^p dr \right\}^{\frac{1}{p}}$$
$$\le \int_0^\infty \left\{ \int_1^N r^{d-1-p\frac{d}{2}+\frac{p}{2}-p\alpha-p} \left| \min\left(rs, \frac{1}{rs}\right) \right|^p dr \right\}^{\frac{1}{p}} |dF_\alpha(s)|.$$

Splitting the integrals, we consider four cases; first we obtain

$$\int_{\frac{1}{N}}^1 |dF_\alpha(s)| \left\{ \int_1^{\frac{1}{s}} r^{d-1-p\frac{d}{2}+\frac{p}{2}-p\alpha-p} (rs)^p\, dr \right\}^{\frac{1}{p}}$$
$$\le C \int_{1/N}^1 s |dF_\alpha(s)| \left(s^{-\frac{d}{p}+\frac{d-1}{2}+\alpha} + 1\right);$$

then

$$\int_0^{\frac{1}{N}} |dF_\alpha(s)| \left\{ \int_1^N r^{d-1-p\frac{d}{2}+\frac{p}{2}-p\alpha-p} (rs)^p\, dr \right\}^{\frac{1}{p}}$$
$$\le C \int_0^{\frac{1}{N}} s |dF_\alpha(s)| \left(N^{\frac{d}{p}-\frac{d-1}{2}-\alpha} + 1\right)$$

and

$$\int_{\frac{1}{N}}^1 |dF_\alpha(s)| \left\{ \int_{\frac{1}{s}}^N r^{d-1-p\frac{d}{2}+\frac{p}{2}-p\alpha-p} (rs)^{-p}\, dr \right\}^{\frac{1}{p}}$$
$$\le C \int_{\frac{1}{N}}^1 \frac{1}{s} |dF_\alpha(s)| \left(N^{\frac{d}{p}-2-\frac{d-1}{2}-\alpha} + s^{-\frac{d}{p}+2+\frac{d-1}{2}+\alpha}\right);$$

and, finally,

$$\int_1^\infty |dF_\alpha(s)| \left\{ \int_1^N r^{d-1-p\frac{d}{2}+\frac{p}{2}-p\alpha-p}(rs)^{-p}\, dr \right\}^{\frac{1}{p}} \le C \int_{\frac{1}{N}}^1 \frac{1}{s} |dF_\alpha(s)| (N^{\frac{d}{p}-2-\frac{d-1}{2}-\alpha} + 1).$$

These give that the third term on the right-hand side of (4.30) may lead to a non-L^p-integrable term if (as opposed to the second term) $\alpha < \frac{d}{p} - \frac{d-1}{2}$.

Remark 4.33. For radial functions with compact support and integrable Fourier transform, Podkorytov [154] obtained a similar formula.

Let us go on to the proof of Theorem 4.29.

Proof. Denoting

$$\Phi_\alpha(r) = \frac{1}{\Gamma(\alpha)} r^{\alpha+\frac{1}{2}} Q_\alpha(r) - \sqrt{\frac{2}{\pi}} \cos\left(r - \frac{\pi}{4}(d + 2\alpha - 1)\right),$$

we then rewrite (4.15) in the form

$$\widehat{f}(x) = C_{1,d}|x|^{-\frac{d}{2}+\frac{1}{2}-\alpha} \int_0^\infty F_\alpha(t) \cos\left(2\pi|x|t - \frac{\pi}{4}(d + 2\alpha - 1)\right) dt + 2\pi(-1)^{\alpha^*+1}|x|^{-\frac{d-1}{2}-\alpha} \int_0^\infty F_\alpha(t)\Phi_\alpha(2\pi|x|t)\, dt,$$

and only the last term has to be discussed. Decomposing it into

$$F_\alpha\left(\frac{1}{|x|}\right) \int_0^\infty \Phi_\alpha(2\pi|x|t)\, dt + \int_0^\infty \left[F_\alpha(t) - F_\alpha\left(\frac{1}{|x|}\right)\right] \Phi_\alpha(2\pi|x|t)\, dt = I_1 + I_2,$$

we have

$$I_1 = \frac{1}{2\pi|x|} F_\alpha\left(\frac{1}{|x|}\right) \int_0^\infty \Phi_\alpha(t)\, dt = C_{2,d}(-1)^{\alpha^*+1} \frac{1}{|x|} F_\alpha\left(\frac{1}{|x|}\right) \tag{4.34}$$

provided $\int_0^\infty \Phi_\alpha(t)\, dt$ is finite. Let us see that this is the case. The integral is obviously convergent over $[0, 1]$. Thus it remains to estimate $\Phi_\alpha(t)$ on $[1, \infty)$. By Lemma 4.21 we have

$$Q_\alpha(t) = \int_0^1 (1 - s)^{\alpha-1} s^{\frac{d}{2}} J_{\frac{d}{2}-1}(ts)\, ds = \Gamma(\alpha)t^{-\alpha} J_{\frac{d}{2}+\alpha-1}(t) + (\alpha - 1)\Gamma(\alpha + 1)t^{-\alpha-1} J_{\frac{d}{2}+\alpha}(t) + O\big(t^{-\alpha-\frac{5}{2}}\big).$$

Using also (4.19), we obtain

$$\Phi_\alpha(t) = 2^{-\frac{5}{2}}\pi^{-\frac{1}{2}}\left[1-4(n/2+\alpha-1)^2\right]\frac{\sin(t-\frac{\pi}{4}(n+2\alpha-1))}{t} + \frac{(\alpha-1)\Gamma(\alpha+1)}{\Gamma(\alpha)}\sqrt{\frac{2}{\pi}}\frac{\cos\left(t-\frac{\pi}{4}(n+2\alpha-1)\right)}{t} + O\left(\frac{1}{t^2}\right),$$

and the integral in question is convergent.

Splitting I_2 into two parts, we have

$$\begin{aligned}
&\int_0^{\frac{1}{|x|}}\left[F_\alpha(t)-F_\alpha\left(\frac{1}{|x|}\right)\right]\Phi_\alpha(2\pi|x|t)\,dt \\
&= -\int_0^{\frac{1}{|x|}}\Phi_\alpha(2\pi|x|t)\int_t^{\frac{1}{|x|}}dF_\alpha(s)\,dt = -\int_0^{\frac{1}{|x|}}\int_0^s\Phi_\alpha(2\pi|x|t)\,dt\,dF_\alpha(s) \\
&= O\Big(\int_0^{\frac{1}{|x|}}\int_0^s[(|x|t)^{\alpha+1/2}(|x|t)^{\frac{d}{2}-1}+1]\,dt\,|dF_\alpha(s)|\Big) \\
&= O\Big(\int_0^{\frac{1}{|x|}}s\,|dF_\alpha(s)|\Big) = \frac{1}{|x|}O\Big(\int_0^{\frac{1}{|x|}}|x|\,s\,|dF_\alpha(s)|\Big).
\end{aligned}$$

Further,

$$\begin{aligned}
&\int_{\frac{1}{|x|}}^\infty\left[F_\alpha(t)-F_\alpha\left(\frac{1}{|x|}\right)\right]\Phi_\alpha(2\pi|x|t)\,dt = \int_{\frac{1}{|x|}}^\infty\Phi_\alpha(2\pi|x|t)\int_{\frac{1}{|x|}}^t dF_\alpha(s)\,dt \\
&= \int_{\frac{1}{|x|}}^\infty\int_s^\infty\Phi_\alpha(2\pi|x|t)\,dt\,dF_\alpha(s) = \frac{1}{|x|}\int_{\frac{1}{|x|}}^\infty\int_{2\pi|x|s}^\infty\Phi_\alpha(t)\,dt\,dF_\alpha(s),
\end{aligned}$$

and finally, by the above asymptotic for Φ_α,

$$\begin{aligned}
&\int_{2\pi|x|s}^\infty\Phi_\alpha(t)\,dt \\
&= 2^{-\frac{5}{2}}\pi^{-\frac{1}{2}}\left[1-4\left(\frac{d}{2}+\alpha-1\right)^2\right]\int_{2\pi|x|s}^\infty\frac{\sin(t-\frac{\pi}{4}(d+2\alpha-1))}{t}\,dt \\
&\quad+\frac{(\alpha-1)\Gamma(\alpha+1)}{\Gamma(\alpha)}\sqrt{\frac{2}{\pi}}\int_{2\pi|x|s}^\infty\frac{\cos(t-\frac{\pi}{4}(d+2\alpha-1))}{t}\,dt \\
&\quad+O\Big(\int_{2\pi|x|s}^\infty\frac{1}{t^2}\,dt\Big) = O\Big(\frac{1}{|x|s}\Big).
\end{aligned}$$

This completes the proof of the theorem. □

4.5 Certain applications

First, we are now able to generalize Theorem 3.34 (or, more precisely, Theorem 3.37) to the radial case. Since it is essentially the L^1-theorem, we restrict ourselves to the case $\alpha = \frac{d-1}{2}$. For the sake of simplicity, let us again use the notation

$$F(t) := F_{\frac{d-1}{2}}(t) = t^{\frac{d-1}{2}} f_0^{(\frac{d-1}{2})}(t).$$

Theorem 4.35. *Let $f_0 \in MV_{\frac{d+1}{2}}$; assume additionally F to be locally absolutely continuous and $F' \in H_{BT}$. Then, for $|x| > 0$,*

$$\widehat{f}(x) = |x|^{-d}\left[C_{1,d} \sin \frac{\pi(d-1)}{2} + C_{2,d}\right] F\left(\frac{1}{4|x|}\right) + |x|^{1-d}\gamma(|x|), \qquad |x| > 0,$$

where $C_{1,d}$ and $C_{2,d}$ are as in Theorem 4.29, while γ is as in Theorem 3.34.

Proof. First of all, the formulas from Theorems 3.34 and 3.37 may be rewritten in a general form for functions defined on the half-axis $(0, \infty)$ as follows:

$$\int_0^\infty \lambda(t) \cos(2\pi rt + \mu)\, dt = \frac{1}{2\pi r} \sin \mu \lambda\left(\frac{1}{4|r|}\right) + \gamma(r).$$

Now, this and (4.30) proves Theorem 4.35. One has only to recall that in (4.30) the third term is integrable for $\alpha = \frac{d-1}{2}$ (see Remark 4.31). □

Let us apply Theorem 4.35. The following two examples were considered in [136].

Example 4.36. Let $d = 3$ and $f = g_1$ be given by (4.10). Observe that $F(t) = tf'(t) \in A_q$, $q > 1$, and

$$\|\gamma\|_{L^1(\mathbb{R})} \le \|F'\|_{H_{BT}} \le C_q \|F'\|_{A_q},$$

thus, by (3.29), the hypotheses of Theorem 4.35 are satisfied. Hence

$$\widehat{f}(x) = \frac{C}{|x|^3 \ln(|x|) \cos(\ln \ln(|x|))} + \frac{1}{|x|^2}\gamma(|x|),\ |x| > 1, \quad x \in \mathbb{R}^3,$$

where

$$\int_1^\infty |\gamma(t)|\, dt < \infty.$$

This is indeed an asymptotic formula since the second term on the right-hand side is integrable outside the ball with the radius 1 unlike the first term which is not integrable! This can be extended at once to the other dimensions when using (4.11) with $\delta = \frac{1}{2}$.

Example 4.37. Let us consider the linear means (1.18) of the Fourier series of 1-periodic in each variable functions $f \in L^1(\mathbb{T}^d)$.

The following theorem is known for the L^1-norms of these means (see, *e.g.*, [125].

Theorem 4.38. *Let $\varphi(x) = \varphi_0(|x|)$ be a radial function such that $\varphi_0 \in MV_{\frac{d+1}{2}}$ and continuous at the origin. Then*

$$\|L_N^\varphi\|_{L^1(\mathbb{T}^d)} = \int_{|x|\le N} |\widehat{\varphi}(x)|\,dx + O(\|F\|_{BV[0,\infty)})$$

where, like above, $F(t) = t^{\frac{d-1}{2}}\varphi_0^{(\frac{d-1}{2})}(t)$.

Let $d = 3$ and

$$\varphi_0(t) = (1 - g_2(t))\chi(t),$$

where χ and g_2 are described in (4.11). Note that we have $\varphi_0(0) = 1$ which is necessary for approximation. Applying Theorem 4.38 and then Theorem 4.35, we obtain

$$\begin{aligned}\|L_N^\varphi\|_{L^1(\mathbb{T}^d)} &= C\int_{1\le|x|\le N} \frac{|\cos(\ln\ln(\frac{2|x|}{\pi}))|}{|x|^d \ln\left(\frac{2|x|}{\pi}\right)\ln\ln\left(\frac{2|x|}{\pi}\right)}\,dx + O(1)\\ &= C\int_1^N \frac{|\cos(\ln\ln r)|}{r\ln r\ln\ln r}\,dr + O(1) = C\ln\ln\ln N + O(1).\end{aligned}$$

Replacing $\ln\ln\left(\frac{1}{t}\right)$ by a "longer" ln-chain in the denominator of the function g_2 one can get a worse behavior in N of $\|L_N^\varphi\|_{L^1(\mathbb{T}^d)}$.

Let us now obtain a generalization, to the multiple case, of the Zygmund–Bochkarev criterion for the absolute convergence of Fourier series of a function of bounded variation (see [28], [29, Ch. 2, Th. 3.1]). We use the standard notation ω for the modulus of continuity.

Corollary 4.39. *Let a radial function be boundedly supported, satisfy conditions (4.2) and (4.4), and $F(0) = 0$. Then the condition*

$$\sum_{k=1}^\infty \frac{\sqrt{\omega\left(F;\frac{1}{k}\right)}}{k} < \infty \tag{4.40}$$

is sufficient and, on the whole class, necessary for $\widehat{f} \in L^1(\mathbb{R}^d)$.

Proof. Since $|F(t)| \le \omega(F;\frac{1}{k})$ for $t \in (\frac{1}{k+1}, \frac{1}{k}]$, the condition

$$\sum_{k=1}^\infty \frac{|F\left(\frac{1}{k}\right)|}{k} < \infty$$

provides that (4.32) holds. Applying Theorem 4.29, we reduce the problem to the one-dimensional one. True, the absolute convergence of the Fourier series was investigated by Bochkarev, but it is closely connected with the integrability of the Fourier transform due to the following theorem of Trigub (see [196, 197]):

Theorem 4.41. *Let $f(t)$ be a boundedly supported function of one variable, and $f_1(t) = tf(t)$. Then $\widehat{f} \in L^1(\mathbb{R}^1)$ if and only if the functions f and f_1 after periodic extension have absolutely convergent Fourier series.*

To use the negative part, Bochkarev's result, it remains to note that

$$\omega\left(F; \frac{1}{k}\right) \leq C \max\{\omega(F,t)\ ; t\}$$

for $t \in \left(\frac{1}{k+1}, \frac{1}{k}\right]$, and the corollary is proved. □

We are not aware of any other multidimensional generalization of the negative, essential part.

Let us give two more examples.

Example 4.42. Consider the function $f(x) = (1 - |x|^\alpha)_+^\beta$ and establish the integrability of its Fourier transform for $\alpha > 0$ and $\beta > \frac{d-1}{2}$ by means of Corollary 4.39. This is important in problems of summability of Fourier series and multipliers. Conditions (4.2) and (4.3) are evidently satisfied. The same may be said about conditions (4.4) and (4.40) for d odd. For d even to verify (4.4), it suffices to show that for

$$\psi(t) = t^\gamma (a - t^\alpha)_+^{\frac{1}{2}+\varepsilon},$$

with $\varepsilon \geq 0$, $\gamma \geq 0$, the function $t^{\frac{1}{2}} \psi^{(\frac{1}{2})}(t)$ is of bounded variation. We have

$$\psi^{(\frac{1}{2})}(t) = \frac{d}{dt} \int_t^1 (s-t)^{-\frac{1}{2}}\ s^\gamma (1-s^\alpha)^{\frac{1}{2}+\varepsilon}\, ds = \int_t^1 (s-t)^{-\frac{1}{2}} \frac{d}{ds}\left\{s^\gamma (1-s^\alpha)^{\frac{1}{2}+\varepsilon}\right\} ds.$$

This means that the boundedness of variation of the function

$$t^{\frac{1}{2}} \int_t^1 (s-t)^{-\frac{1}{2}}\ s^{\zeta-1} (1-s^\alpha)^{\varepsilon-\frac{1}{2}} ds,$$

with $\varepsilon \geq 0$, $\zeta > 0$, should be established. Further,

$$s^{\zeta-1}(1-s^\alpha)^{\varepsilon-\frac{1}{2}} = (1-s^\alpha)^{\varepsilon-\frac{1}{2}}(s^{\zeta-1}-1) + (1-s^\alpha)^{\varepsilon-\frac{1}{2}}.$$

Denoting $C_\alpha = \lim\limits_{s\to1}\left(\frac{1-s^\alpha}{1-s}\right)^{\varepsilon-\frac{1}{2}}$, we have

$$\begin{aligned}(1-s^\alpha)^{\varepsilon-\frac{1}{2}} &= (1-s^\alpha)^{\varepsilon-\frac{1}{2}} - C_\alpha(1-s)^{\varepsilon-\frac{1}{2}} + C_\alpha(1-s)^{\varepsilon-\frac{1}{2}} \\ &= (1-s^\alpha)^{\varepsilon+\frac{1}{2}} \frac{\left(\frac{1-s^\alpha}{1-s}\right)^{\varepsilon-\frac{1}{2}} - C_\alpha}{1-s} + C_\alpha(1-s)^{\varepsilon-\frac{1}{2}}.\end{aligned}$$

Thus, the boundedness of variation of the function

$$t^{\frac{1}{2}} \int_t^1 (s-t)^{-\frac{1}{2}} (1-s)^{\varepsilon-\frac{1}{2}} \, ds$$

should be established. But, by a simple change of variables, this function is equal to the function

$$t^{\frac{1}{2}}(1-t)^{\varepsilon} \int_0^1 s^{-\frac{1}{2}}(1-s)^{\varepsilon-\frac{1}{2}} \, ds,$$

and (4.4) is now obvious. Moreover, this makes (4.40) obvious as well, which completes the proof of the example.

Example 4.43. Further, consider

$$f(x) = \frac{\left(1-(1-|x|^{\alpha})_+^{\beta}\right)}{|x|^r}$$

with $\alpha > r$ and $\beta > \frac{d-1}{2}$. Let us show again that $\widehat{f} \in L^1(\mathbb{R}^d)$. This is of importance in approximation of a function by linear means (1.18) generated by our f on place of φ there. On $[0,1]$ the argument from Example 4.42 is applicable. Hence for $t \in [0,1]$ we have

$$F(t) = C_1 t^{\alpha-r}(1-t)^{\beta-\frac{d-1}{2}} + g(t),$$

where g is a continuously differentiable function, $g(0) = 0$. It is clear that (4.32) is satisfied. Applying, if needed, the formula of fractional derivation, we obtain $t^{\frac{d-1}{2}} (t^{-r})^{(\frac{d-1}{2})} = C_2 t^{-r}$. Using now Theorem 4.29 and integrating by parts in the one-dimensional integral, we arrive to estimating the following value:

$$\int_1^N \left| \int_0^{\infty} F'(t) \cos\left(st - \frac{\pi d}{2}\right) dt \right| \frac{ds}{s}.$$

On $[0,1]$ we have $F'(t) \in \operatorname{Lip}\varepsilon$ in the L^1 metrics, for some $\varepsilon > 0$. But the integral $\int_1^N s^{-1-\varepsilon} ds$ converges. For $t \in [1,\infty]$, integration by parts yields

$$\begin{aligned}
&\int_1^N \left| \int_1^{\infty} \frac{\cos(st - \pi\frac{d}{2})}{t^{1+r}} \, dt \right| \frac{ds}{s} \\
&= \int_1^N \left| \frac{1}{s} t^{-1-r} \sin\left(st - \pi\frac{d}{2}\right) \Big|_1^{\infty} + \left(\frac{1}{s} + \frac{r}{s}\right) \int_1^{\infty} \frac{\sin(st - \frac{\pi d}{2})}{t^{2+r}} \, dt \right| \frac{ds}{s} \\
&\le C \int_1^N \frac{1}{s^2} \, ds,
\end{aligned}$$

and Example 4.43 is proved.

The examples considered show that Theorem 4.29 covers a wide range of functions important in applications and is friendly enough.

Part 2

Geometric (and Analytic) Aspects

Chapter 5

L^2-average Decay of the Fourier Transform of a Characteristic Function of a Convex Set

Let B be a bounded open set in $\mathbb{R}^d$. As we note in the introduction, it is a consequence of the classical method of stationary phase that if ∂B is sufficiently smooth and has everywhere non-vanishing Gaussian curvature, then

$$|\widehat{\chi}_B(R\omega)| \lesssim R^{-\frac{d+1}{2}}, \tag{5.1}$$

with constants independent of ω. The estimate (5.1) is optimal in a very strong sense. One can check that a better rate of decay at infinity is not possible. One can also check that if the Gaussian curvature vanishes at even a single point, then (5.1) does not hold.

In fact, the pointwise estimate may be much worse. For example, if B is convex, one has

$$|\widehat{\chi}_B(R\omega)| \lesssim R^{-1},$$

and the case of a cube $[0,1]^d$ shows that one cannot, in general, do any better. See, for example, [180], for a nice description of these classical results.

It is perhaps even more surprising that non-vanishing curvature alone is not enough to guarantee that the estimate (5.1) holds if ∂B is not sufficiently smooth. See, for example, [110].

In spite of the fact that the estimate (5.1) does not hold in general, a basic question is whether this estimate holds on average for a large class of domains, for example, bounded open sets with a rectifiable boundary. More precisely, one should like to know for which domains one has the following estimate:

$$\left(\int_{S^{d-1}} |\widehat{\chi}_B(R\omega)|^2 d\omega\right)^{\frac{1}{2}} \lesssim R^{-\frac{d+1}{2}}. \tag{5.2}$$

The next three chapters will be dedicated to this question and its variants. For example, in some cases, it is equally useful to know whether

$$\left(\int_{S^{d-1}} |\widehat{\sigma}(R\omega)|^2 d\omega\right)^{\frac{1}{2}} \lesssim R^{-\frac{d-1}{2}}, \tag{5.3}$$

where σ is the Lebesgue measure on the boundary of B. Under a variety of assumptions, for example, if ∂B is Lipschitz, (5.2) and (5.3) are linked via the divergence theorem. We use this fact in the proof of our main result below.

An example due to Sjölin ([174]) shows that (5.3) is not purely dimensional. He showed that if σ is an arbitrary $(d-1)$-dimensional compactly supported measure, then the best exponent one can expect on the right-hand side of (5.3) is $\frac{d-\frac{3}{2}}{2}$. This means that in order to prove an estimate like (5.2) we must use the fact that ∂B is in some sense a hyper-surface.

Several results of this type have been proved over the years. In [155], Podkorytov proved (5.2) for convex domains in two dimensions using a beautiful geometric argument that relied on the fact that in two dimensions, the Fourier transform of a characteristic function of a convex set in a given direction is bounded by a measure of a certain geometric cap. See, for example, [46] or [41] for more details. Unfortunately, in higher dimensions one cannot bound the Fourier transform of a characteristic function of a convex set by such a geometric quantity. See, for example, [8]. For the case of average decay on manifolds of co-dimension greater than one, see, *e.g.*, [53], [141], and [109].

The analytic case has been known for a long time. See, for example, [160]. In [204], Varchenko proved (5.1) under the assumption that ∂B is sufficiently smooth. Smoothness allows one to use the method of stationary phase in a very direct and strong way. In the general case, one must come to grips with the underlying geometry of the problem. In the main result of this chapter, we drop the smoothness assumption and prove that (5.2) holds for all bounded open convex sets B in $\mathbb{R}^d$. In addition, we prove the same estimate under an assumption that the boundary is $C^{\frac{3}{2}}$.

The main geometric feature of our approach is a quantitative exploitation of the following simple idea: if $\omega \in S^{d-1}$ is normal to ∂B at x, and y is sufficiently close to x, then $x-y$ cannot be parallel to ω. This allows us to deal with the so-called "stationary" points of the oscillatory integral resulting from (5.3). Unlike the smooth case, where "non-stationary" points are very easy to handle using integration by parts, in the general case one is forced to exploit the smoothness of the sphere along with an appropriate integration by parts argument that exploits either convexity or the $C^{\frac{3}{2}}$ assumption on the boundary.

The estimates (5.2) and (5.3) have numerous applications in various problems of harmonic analysis, analytic number theory and geometric measure theory. Moreover, (5.2) and (5.3) imply immediate generalizations of a number of results in analysis and analytic number theory to higher dimensions. See, for example,

[38], [40], [41], [55], [53], [102], [109], [115], [142], [148], [160], [163], [174], [175], and [204]. We give two simple examples to illustrate the point.

The proof is based on (5.1). Using (5.2) instead, one can prove the following version:

$$\left(\int_{S^{d-1}} \left|\#\{t\rho B \cap \mathbb{Z}^d\} - t^d|B|\right|^2 d\rho\right)^{\frac{1}{2}} \leq Ct^{d-2+\frac{2}{d+1}}, \tag{5.4}$$

where ρB denotes the rotation of B by $\rho \in S^{d-1}$ viewed as an element of $SO(d)$. See, for example, [107] and [44]for a detailed discussion of applications of average decay of the Fourier transform to lattice point problems. Also note that Theorem 5.7 below shows that convexity may be replaced by a $C^{\frac{3}{2}}$ assumption. The extent to which the $C^{\frac{3}{2}}$ assumption is sharp is still an open question.

5.1 L^2-average decay

Our main results are the following two theorems.

Theorem 5.5. *Let B be a bounded convex domain in $\mathbb{R}^d$. Then*

$$\int_{S^{d-1}} |\widehat{\chi}_B(R\omega)|^2 d\omega \lesssim R^{-(d+1)}. \tag{5.6}$$

Theorem 5.7. *Let B be an bounded open set in $\mathbb{R}^d$ satisfying the following assumption. The boundary of B can be decomposed into finitely many neighborhoods such that given any pair of points P, Q in the neighborhood,*

$$|(P-Q)\cdot n(Q)| \lesssim |P-Q|^{\frac{3}{2}}, \tag{5.8}$$

where $n(Q)$ denotes the unit normal to ∂B at Q. Then (5.1) *holds.*

The first result completely settles the question of average decay for convex sets with convex boundaries and immediately raises the question of whether one can reasonably go beyond convexity. The second result is certainly a significant step in that direction as the condition essentially amounts to the $C^{\frac{3}{2}}$ assumption on the boundary of the set. However, we do not see any reason why the result should not hold for compact sets with Lipschitz boundaries. In other words, there is much room between the conditions under which we can prove our results and the counter-examples due to Sjölin mentioned above.

5.2 Proof of Theorem 5.5 and Theorem 5.7

We shall give simultaneous proofs of Theorem 1.1 and Theorem 1.2. The argument is based on the fact that both convex surfaces and $C^{\frac{3}{2}}$ surfaces satisfy the following geometric condition:

The boundary of B can be decomposed into finitely many neighborhoods B_j such that on each neighborhood the surface is given as a graph of a Lipschitz function with the Lipschitz constant < 1.

The geometric meaning of this condition is that for x and y belonging to the same neighborhood B_j, the secant vector $x - y$ lies strictly within $\frac{\pi}{4}$ of our local coordinate system's horizon. We shall henceforth refer to this as the "secant property".

This geometric condition is clearly satisfied by C^1 (and hence $C^{\frac{3}{2}}$) surfaces, by taking the neighborhoods to be sufficiently small. For convex domains, we make the following construction. We cover S^{d-1} by a smooth partition of unity η_j, such that $\sum_j \eta_j \equiv 1$ and such that support of each η_j is contained in the intersection of the sphere and a cone of aperture strictly smaller than $\frac{\pi}{2}$. Then if $n(x)$ denotes the Gauss map taking $x \in \partial B$ to the unit normal at x, then $\sum_j \eta_j(n(x))$ induces the desired decomposition on the boundary of B. We note that in the convex case the number of such neighborhoods depends only on dimension.

By the divergence theorem,

$$\widehat{\chi}_B(R\omega) = -\frac{1}{2\pi i R} \int_{\partial B} e^{-ix\cdot R\omega} \left(\omega \cdot n(x)\right) d\sigma(x), \tag{5.9}$$

where $n(x)$ denotes the unit normal to ∂B at x, and $d\sigma$ denotes the surface measure on the boundary. This reduces the problem to the boundary of B.

5.2.1 Decomposition of the boundary

Let ϕ_j denote a smooth partition of unity on ∂B subordinate to the decomposition

$$\partial B = \cup_{j=1}^{N} B_j.$$

Moreover, ϕ_js are chosen such that on the support of each ϕ_j, the aforementioned secant property still holds. It follows that the corresponding Lipschitz constant K_j is less than 1. This is the basic building block of our proof.

Let ψ_j be a smooth cutoff function identically equal to 1 on the spherical cap of solid angle $> \frac{\pi}{2}$ and which is supported in a slightly bigger spherical cap which lies at a strict positive distance from all the vectors $\frac{x-y}{|x-y|}$, $x, y \in supp(\phi_j) \subset \partial B$. Notice that our hypothesis makes such a decomposition possible and that all the vectors normal to ∂B on the support of ϕ_j lie strictly inside the support of ψ_j.

5.2.2 Singular directions

This part of the proof is identical in the convex and the $C^{\frac{3}{2}}$ cases. In fact, it depends only on the secant property. Let

$$F_j(R\omega) = \int_{\partial B} e^{-ix\cdot R\omega} d\mu_j(x),$$

where

$$d\mu_j = (\omega \cdot n(x))\, \phi_j(x) d\sigma(x).$$

In view of (5.9) and the triangle inequality, it suffices to show that

$$\int_{S^{d-1}} |F_j(R\omega)|^2 d\omega \lesssim R^{-(d-1)}.$$

Now,

$$\int_{S^{d-1}} |F_j(R\omega)|^2 d\omega$$
$$= \int_{S^{d-1}} |F_j(R\omega)|^2 \psi_j(\omega) d\omega + \int_{S^{d-1}} |F_j(R\omega)|^2 (1-\psi_j(\omega)) d\omega = I + II.$$

We shall refer to the support of ψ_j as "singular" directions, and the other vectors on the sphere as "non-singular" directions. The origin of this notation is the fact that in the smooth case, the singular, or stationary directions are the ones that are normal to the relevant piece of the hyper-surface in question.

We have

$$I = \int_{\partial B} \int_{\partial B} \int_{S^{d-1}} e^{i(x-y)\cdot R\omega} \psi_j(\omega) d\mu_j(x) d\mu_j(y).$$

Using the definition of ψ_j, we integrate by parts N times and obtain

$$I \lesssim \int_{\partial B} \int_{\partial B} \min\{1, (R|x-y|)^{-N}\} d\mu_j(x) d\mu_j(y) \lesssim R^{-(d-1)},$$

since $d\mu_j$ is $d-1$-dimensional and compactly supported.

5.2.3 Non-singular directions

We shall take the following perspective on the spherical coordinates. Let $\omega = \omega(\tau_1, \dots, \tau_{d-2}, \theta)$, where $(\tau_1, \dots, \tau_{d-2})$ denotes the "azimuthal" angles, and θ denotes the remaining angle, *i.e.*, $\theta = \tan^{-1}(\frac{x_d}{x_1})$. Note that for each fixed θ, $(\tau_1, \dots, \tau_{d-2})$ give a coordinate system on the "great circle" tilted at the angle θ from the horizontal.

For each fixed θ, we set up a coordinate system such that

$$II = \int_{-\frac{\pi}{2}}^{\frac{\pi}{2}} \int \left| \int e^{iR\omega' \cdot u} \Phi_\theta(u) du \right|^2 J(\tau, \theta)(1-\psi_j(\omega)) d\tau d\theta, \qquad (5.10)$$

where $\omega = \omega(\tau, \theta)$, $\omega = (\omega', \omega_d)$,

$$\Phi_\theta(u) = \phi_j(u, A_\theta(u)) \sqrt{1 + |\nabla A_\theta(u)|^2},$$

and J is the (smooth) Jacobian corresponding to the spherical coordinates.

Here we are viewing this portion of the boundary of B as the graph, of the function A_θ, above the hyperplane determined by the $(d-2)$-dimensional "great circle" obtained by fixing θ.

By a further partition of unity, a rotation, and the triangle inequality, we may assume that we are in an arbitrarily small neighborhood of $\omega = (1, 0, \ldots, 0)$.

The key object in the remaining part of the proof is the difference operator

$$\Delta_h f(s) = f(s+h) - f(s).$$

We observe that the transpose of this operator

$$\Delta_h^* = \Delta_{-h}.$$

We also note that

$$\Delta_{\frac{1}{R}}(e^{iR\omega_1 u_1}) = (e^{i\omega_1} - 1)e^{iR\omega_1 u_1}.$$

Then by discrete integration by parts, the square root of the portion of (5.10) in the neighborhood of $(1, 0, \ldots, 0)$ equals

$$\left(\int_{-\frac{\pi}{2}}^{\frac{\pi}{2}} \int \left| \frac{1}{e^{i\omega_1} - 1} \int \int e^{iR\omega' \cdot u} \Delta_{-\frac{1}{R}} \Phi_\theta(\cdot, u')(u_1) du_1 du' \right|^2 J(\tau, \theta) \Psi_j(\omega) d\tau d\theta \right)^{\frac{1}{2}}, \tag{5.11}$$

where Ψ_j is an appropriate cut-off function supported in the neighborhood of $(1, 0, \ldots, 0)$, and $u' = (u_2, \ldots, u_{d-1})$.

Applying the Minkowski integral inequality, we see that (5.11) is bounded by

$$\int \left(\int_{-\frac{\pi}{2}}^{\frac{\pi}{2}} \int \left| \int e^{iR\omega'' \cdot u'} \Delta_{-\frac{1}{R}} \Phi_\theta(\cdot, u')(u_1) du' \right|^2 J(\tau, \theta) \Psi_j(\omega) d\tau d\theta \right)^{\frac{1}{2}} du_1, \tag{5.12}$$

where $\omega'' = (\omega_2, \ldots, \omega_{d-1})$. For a fixed θ, the integration in τ is over the $(d-2)$-dimensional "great circle". We may parameterize the sphere so that this "great circle" is given by $\omega_1 = \omega_1(\omega'')$.

Expanding (5.12) and rewriting, we get

$$\begin{aligned} \int \Bigg(\int_{-\frac{\pi}{2}}^{\frac{\pi}{2}} \int \int e^{iR\omega'' \cdot (u' - v')} \Delta_{-\frac{1}{R}} \Phi_\theta(\cdot, u')(u_1) \Delta_{-\frac{1}{R}} \Phi_\theta(\cdot, v')(u_1) du' dv' \\ J'(\omega'', \theta) \Psi_j(\omega) d\omega'' d\theta \Bigg)^{\frac{1}{2}} du_1, \end{aligned} \tag{5.13}$$

where $J'(\omega'', \theta)$ is smooth in ω''.

Integrating by parts in ω'' we see that (5.13) is bounded by

$$\int \Big(\int_{-\frac{\pi}{2}}^{\frac{\pi}{2}} \int \int \min\{1, R|u'-v'|^{-N}\} |\Delta_{-\frac{1}{R}} \Phi_\theta(\cdot,u')(u_1)|$$

$$|\Delta_{-\frac{1}{R}} \Phi_\theta(\cdot,v')(u_1)| du' dv' q(\theta) d\theta \Big)^{\frac{1}{2}} du_1 \tag{5.14}$$

$$\lesssim R^{-\frac{d-2}{2}} \int \left(\int_{-\frac{\pi}{2}}^{\frac{\pi}{2}} \int |\mathcal{M}' \Delta_{-\frac{1}{R}} \Phi_\theta(\cdot,u')(u_1)|^2 du' q(\theta) d\theta \right)^{\frac{1}{2}} du_1,$$

where $\mathcal{M}'$ is the Hardy–Littlewood maximal function in the u' variable, and q is a smooth cut-off function. The last inequality uses the standard fact that convolution with a radial, integrable and decreasing kernel is dominated by the maximal function. See, for example, [179, Chapter 3]. Since our integrand is compactly supported in the u_1 variable, we may apply Bunyakovskii–Cauchy–Schwarz to see that the right-hand side of (5.14) is bounded by

$$R^{-\frac{d-2}{2}} \left(\int_{-\frac{\pi}{2}}^{\frac{\pi}{2}} \int \int |\Delta_{-\frac{1}{R}} \Phi_\theta(\cdot,u')(u_1)|^2 du' du_1 q(\theta) d\theta \right)^{\frac{1}{2}},$$

since the Hardy–Littlewood maximal function is bounded on L^2.

The conclusion of the theorem now follows from the estimate

$$\|\Delta_{\frac{1}{R}} \Phi_\theta\|_{L^2(du)} \leq CR^{-\frac{1}{2}}. \tag{5.15}$$

Clearly (5.15) holds if $\partial B \in C^{\frac{3}{2}}$, for in that case $\Phi_\theta \in C^{\frac{1}{2}}$ with compact support. In the convex case we interpolate between the estimates

$$\|\Delta_{\frac{1}{R}} \Phi_\theta\|_{L^\infty(du)} \leq C, \tag{5.16}$$

and

$$\|\Delta_{\frac{1}{R}} \Phi_\theta\|_{L^1(du)} \leq CR^{-1}, \tag{5.17}$$

where (5.16) holds because convex surfaces are Lipschitz (hence Φ_θ is bounded), and (5.17) holds by the mean-value theorem, Fubini's theorem, and Gauss–Bonnet's theorem (for cross-sections) in the u_1 variable. □

Chapter 6

L^1-average Decay of the Fourier Transform of a Characteristic Function of a Convex Set

In the previous chapter we obtained optimal L^2-average decay under the assumption that the set B is bounded and has a convex boundary. We now turn our attention to obtaining more detailed understanding of this problem in the two-dimensional setting.

6.1 Preliminary discussion

Roughly speaking, we now understand that on L^2, "all cats are grey in the dark" in the sense that there is no difference between, say, a cube and the ball as far as the average decay rate of the Fourier transform of the characteristic function goes. We know from the classical stationary phase that the pointwise decay (L^∞ decay depends on smoothness and curvature properties of the boundary. This raises a compelling question of whether interesting ideas arise in the analysis of L^p-average decay where $1 \leq p < 2$. To a significant extent this is the subject of this chapter.

Part of the motivation for the topics in this chapter come from the fact that many classical problems in analysis, geometry, and number theory are stated in terms of basic properties of such sets. For example, we may consider the difference between the number of lattice points inside the dilated set ρB and its area, *i.e.*, the *discrepancy*

$$D_\rho(B) = \mathrm{card}\left(\rho B \cap \mathbb{Z}^2\right) - \rho^2 \left|B\right|$$

where $|\cdot|$ denotes the area. Among the many natural questions we can ask about this problem (see the section on lattice points below) is, how does the geometry of B affect the growth rate of the discrepancy function? As we shall see, there are results that do not distinguish among various convex sets. However, we shall also see that the behavior of the above discrepancy functions corresponding to different convex sets may vary dramatically, and that this behavior may be described in terms of natural and readily computable geometric quantities.

The above question on lattice points has a consequence in the study of irregularities of distribution. Suppose $\mathcal{P} = \{z_j\}_{j=1}^{N}$ is a distribution of N points in the unit square $U = [0,1]^2$ treated as the torus $\mathbb{T}^2$. Let B be a convex body in U with diameter smaller than 1. Assume $\varepsilon \leq 1$, $t \in \mathbb{T}^2$. Then certain sharp upper estimates for the discrepancy

$$D(\mathcal{P}, \varepsilon, t) = \sum_{j=1}^{N} \chi_{\varepsilon B - t}(z_j) - N\,\varepsilon^2\,|B|$$

can be obtained from related estimates for lattice points (by a suitable trick we shall reduce to the case when N is a square, which in turns is an easy corollary).

At the heart of the lattice point and the irregularities of distribution problems is the Fourier transform of the characteristic function of B. Our approach is to study the effect of the geometric properties of B on the decay rate of the Fourier transform of the characteristic function of B and its variants. We shall then use this analysis to obtain precise information about the discrepancy functions described above.

How should we distinguish among the various convex planar sets? The lattice point problem suggests one natural approach. It was observed by Gauss that $D_\rho(B) \lesssim \rho$, since the boundary of B is one-dimensional. Consider the case when B is a unit square with sides parallel to the axis. When ρ is an integer, the boundary of ρB contains $\approx \rho$ integer lattice points, thus showing that this estimate cannot be improved. However, if B is a disc, the boundary of ρB "curves away" from the integer lattice. In fact, it is known (see [116]) that the estimate for $D_\rho(B)$ in this case is much better. These two examples suggest that the curvature of the boundary may be the key distinguishing factor among convex sets. The boundary of the square has no curvature, which leads to a poor discrepancy estimate, where the boundary of the disc has everywhere non-vanishing curvature, and the estimate for the discrepancy function is considerably better.

The notion of curvature alluded to in the previous paragraph is the standard geometric, or Gaussian, curvature, defined as the determinant of the differential of the Gauss map which maps each point on the boundary of a convex set to the unit normal at that point. It turns out that the geometric curvature alone does not capture the relevant properties of convex planar sets fully. To see this, let us return to the case of the unit square. While it is true that the discrepancy function is terrible if the sides of the square are parallel to the axes, the discrepancy function becomes practically non-existent, even better than the discrepancy function for the disc, if the square is rotated by a sufficiently irrational angle (see [62]). In fact, it is precisely the "flatness" of the squares that keeps its boundary from hitting hardly any lattice points when it is rotated. This suggests that for "most" rotations, convex sets with "flat" boundaries behave better as far as discrepancy functions are concerned.

In this chapter we consider the rotated and translated copies

$$\sigma^{-1}(\rho B) - t,$$

where $\sigma \in SO(2)$, $t \in \mathbb{T}^2$, of the dilated body ρB (here ρ is a large positive number) and we study the L^1 mean

$$\int_{\mathbb{T}^2} \int_{SO(2)} \left| D_\rho(\sigma^{-1}(B) - t) \right| d\sigma$$

of the discrepancy

$$D_\rho(\sigma^{-1}(B) - t) = \operatorname{card}\left((\rho\sigma^{-1}(B) - t) \cap \mathbb{Z}^2 \right) - \rho^2 |B| .$$

The reason for choosing the L^1 mean among other L^p means will be clear soon. Let us also say that in many cases, averaging makes a discrepancy problem easier. For example, the Gauss circle problem is a basic and unsolved problem, while one can obtain (see, *e.g.*, [114] or [41]) a sharp result averaging in L^2 over translations of the discs and using only Parseval's identity and some properties of Bessel functions.

Let us go back to the geometry of B. The above observations can be exploited in a number of ways. If "flatness" is good, then B, *i.e.*, the family of rotated copies of B, is better if B is close to being a polygon. This means that B is good if it can be approximated by a polygon with relatively few sides (the construction we are going to describe has been studied in [155] and [171], see also [210]). We choose an arbitrary point on the boundary of B and draw a chord to another point on the boundary of B in such a way that the maximum distance from the chord to the boundary of B is ρ^{-1}. Roughly speaking, if the number of sides of the above inscribed polygon is $\lesssim \rho^\alpha$, we say that the dimension of B is at least α (we shall explain later why for most of the chapter we prefer not to consider the infimum of the α's). Note that B is a polygon if and only if we can choose $\alpha = 0$, and if B is a circle then, $\alpha = \frac{1}{2}$ works.

We can also take the following "dual" point of view. If B is close to a polygon, then its boundary ∂B has relatively few normals. A more precise way of saying this is that the area of the δ-neighborhood of the image of ∂B under the Gauss map is $\lesssim \delta^{1-d}$. If B is a disc, we can only take $d = 1$. On the other hand, we can choose $d = 0$ if and only if B is a polygon. As another example, let B be a polygon with infinitely many sides the normals of which have apertures in the sequence $n^{-\beta}$, $\beta > 0$, it is easy to see that in this case we can take $d = (1+\beta)^{-1}$.

Introducing the infima α^* and d^* (note that d^* is the upper Minkowski dimension of the image of the Gauss map) we have

$$\alpha^* \leq \frac{d^*}{d^* + 1}$$

and we can also prove that this bound is best possible. On the other hand we can show that α^* can be as close to 0 as we want, even when d^* is away from 0.

We conclude this introductory part by noting that a notion of a dimension of a convex set may be applicable and natural in a number of interesting problems in analysis and combinatorics. For example, the Falconer distance conjecture says that if the Hausdorff dimension of a planar set is greater than 1, then the set of Euclidean distances among the points of this set has positive Lebesgue measure. However, if the Euclidean distance is replaced by the "taxi-cab" (l^1) metric, the conjecture is clearly false, and in fact the set is required to have Hausdorff dimension 2 before the same conclusion on the distance set possible. It is reasonable to ask whether distances induced by convex sets with "intermediate dimension" provide examples of intermediate behavior in the Falconer Distance Problem.

6.2 A variety of arguments

We are going to discuss various aspects of the problem this chapter deals with.

6.2.1 L^p-average decay of the Fourier transform

The study of the decay of the Fourier transform

$$\widehat{\chi}_B(\xi) = \int_B e^{-2\pi i\xi\cdot x}dx$$

as $|\xi| \to \infty$ is a classical subject. We have already discussed certain aspects of this problem in Section 2.3 of Chapter 2. When ∂B has strictly positive curvature, then $|\widehat{\chi}_B(\xi)| \lesssim |\xi|^{-\frac{3}{2}}$. However, when ∂B contains points where the Gaussian curvature vanishes, the above inequality is no longer true. For example, when B is a polygon and $\Theta = (\cos\theta.\sin\theta)$, then $\widehat{\chi}_B(\rho\Theta)$ decays as ρ^{-1} in some directions and as ρ^{-2} in most directions. In such cases it is useful to study the L^p spherical average decay of $\widehat{\chi}_B$, given by

$$\|\widehat{\chi}_B(\rho\cdot)\|_{L^p(\Sigma_1)} \tag{6.1}$$

where Σ_1 is the unit circle and $1 \le p \le \infty$. Here a basic result is Podkorytov's theorem

$$\|\widehat{\chi}_B(\rho\cdot)\|_{L^2(\Sigma_1)} \lesssim \rho^{-\frac{3}{2}}, \tag{6.2}$$

(see [155]) where no regularity assumption on the boundary ∂B is required.

Throughout this chapter $X \lesssim Y$ will mean, as above, that $X \le cY$, with c depending here only on the body B under consideration. Moreover we shall always assume $\rho \ge 2$.

The study of (6.1) turns out to have applications to several problems, such as the distribution of lattice points in large convex domains ([160], [187], [40], [41]), irregularities of distribution ([141], [40]), summations of multiple Fourier expansions ([42], [38], [39]), and estimates for generalized Radon transforms ([163]).

The paper [41] contains the following rather complete study of (6.1) under the additional assumption that ∂B is piecewise smooth. When $p = 2$, (6.2) says that the rate of decay of (6.1) is independent of the shape of B. When $2 < p \leq \infty$, any order of decay between the one of the disc and the one of the polygon is possible. On the other hand, when $1 \leq p < 2$, a convex body with piecewise smooth boundary behaves either like a disc or like a polygon. In particular, when P is a polygon we have the sharp bound

$$\|\widehat{\chi}_P(\rho\cdot)\|_{L^1(\Sigma_1)} \lesssim \rho^{-2}\log\rho, \tag{6.3}$$

and when B has piecewise smooth boundary, but it is not a polygon, we have the sharp bound

$$\|\widehat{\chi}_B(\rho\cdot)\|_{L^1(\Sigma_1)} \lesssim c\rho^{-\frac{3}{2}}. \tag{6.4}$$

Actually, (6.4) is sharp whenever ∂B contains at least one point where the Gaussian curvature exists and is different from zero.

The above dichotomy pointed out in [41] is no longer valid for arbitrary convex bodies. The existence of "chaotic" decays has been pointed out in [41, p. 553] using an abstract argument on convex sets. Unfortunately, that argument is not constructive, nor does it provide non-trivial explicit bounds for the average decay.

The main analytic tool of this chapter is the L^p-average decay for arbitrary convex planar bodies when $1 \leq p \leq 2$. In essence, we shall consider the L^1-average decay and the L^2-average decay. The results for intermediate exponents can be essentially obtained by interpolation. Roughly speaking, the L^2-average decay is a "all cats are grey in the dark" phenomenon, where the decay does not distinguish among the different convex bodies. On the other hand, the L^1-average decay determines, in a sense, how close a convex set is to a polygon.

6.2.2 Inscribed polygons

We introduce the following notation. For any $\Theta = (\cos\theta, \sin\theta)$ and any small $\delta > 0$ let

$$\begin{aligned} K_\theta &= \max_{x\in B} x\cdot\Theta, \\ r(B,\delta,\theta) &= \{y \in B : y\cdot\Theta = K_\theta - \delta\}. \end{aligned} \tag{6.5}$$

We say that the chord $r(B,\delta,\theta)$ is of height δ and we use it to define the following inscribed polygon (see also [155] or [171]).

Definition 6.6. Let B be a convex planar body. Choose any chord of height δ and name it ch_1. Move counterclockwise constructing a finite sequence of consecutive chords of height δ until you reach ch_1. Then, if necessary, replace the last chord by one consecutive to ch_1 (hence of height not greater than δ). In this way we get a polygon inscribed in B and we denote it by P_δ^B. Of course P_δ^B depends on the choice of ch_1 and we should write $P_\delta^B(ch_1)$, however, none of our results depends on ch_1 and, by a small abuse, we shall always speak about "the" inscribed polygon P_δ^B. We denote by M_δ^B be the number of sides of P_δ^B.

It has been proved in [171] that $M_\delta^B \lesssim \delta^{-\frac{1}{2}}$. Our first result is the following.

Theorem 6.7. *Let B be a convex planar body and assume $M_{\rho^{-1}}^B \lesssim \rho^\alpha$ (where $0 < \alpha < \frac{1}{2}$, the cases $\alpha = 0$ and $\alpha = \frac{1}{2}$ being covered by (6.3) and (6.2) respectively). Then*

$$\|\widehat{\chi}_B(\rho\cdot)\|_{L^1(\Sigma_1)} \lesssim \rho^{\alpha-2} \log \rho. \tag{6.8}$$

Moreover, for any $0 < \alpha < \frac{1}{2}$, there exists a convex planar body B such that $M_{\rho^{-1}}^B \lesssim \rho^\alpha$ and, for any $\varepsilon > 0$,

$$\limsup_{\rho\to+\infty} \rho^{-\alpha+2+\varepsilon} \|\widehat{\chi}_B(\rho\cdot)\|_{L^1(\Sigma_1)} > 0.$$

All the proofs will be given in the last section of this chapter.

Before going on, we want to discuss the above theorem. The first step in the proof is to show that

$$\int_0^{2\pi} |\widehat{\chi}_B(\rho\Theta)| \, d\theta \lesssim \int_0^{2\pi} \left|\widehat{\chi}_{P_{\rho^{-1}}^B}(\rho\Theta)\right| d\theta.$$

(see Definition 6.6). We are therefore reduced to estimating the average decay for a polygon with $\lesssim \rho^\alpha$ sides. The second step simply consists in recalling that the implicit constant in (6.3) depends on the number of sides of the polygon P, and that after reading the proofs in [40] or [41] one can rewrite (6.3) in the following way,

$$\int_0^{2\pi} |\widehat{\chi}_P(\rho\Theta)| \, d\theta \le cN\rho^{-2} \log \rho \tag{6.9}$$

where N is the number of sides of the polygon P, and the constant c is absolute (there is no loss of generality assuming that the length of the boundary ∂P is ≤ 1). Putting ρ^α in place of N we then get (6.8).

At this point one should expect to have gotten a *poor* result using the trivial estimate (6.9). The counterexample in the theorem shows that it is not so.

6.2.3 The image of the Gauss map

At every point of ∂B there is a left and a right tangent, therefore a left ($-$) and a right ($+$) outward normal. Let $\pi^\pm : \partial B \to \Sigma_1$ be the map sending each point in ∂B to the left/right normal. Also let

$$\Delta^B = \pi^-(\partial B) \cup \pi^+(\partial B). \tag{6.10}$$

We identify Σ_1 with the interval $[0, 2\pi)$. For every $\theta \in [0, 2\pi)$ we denote with $d(\theta, \Delta^B)$ the distance between θ and Δ^B. For a given small δ, let

$$\Delta_\delta^B = \left\{x \in [0, 2\pi) : d(x, \Delta^B) < \delta\right\} \tag{6.11}$$

be the δ-neighborhood of Δ^B.

Theorem 6.12. *Let* $0 < d < 1$. *Assume*

$$\left|\Delta_\delta^B\right| \lesssim \delta^{1-d}, \tag{6.13}$$

then

$$\|\widehat{\chi}_B(\rho\cdot)\|_{L^1(\Sigma_1)} \lesssim \rho^{\frac{d}{d+1}-2}. \tag{6.14}$$

Moreover there exists a convex body B *satisfying* $\left|\Delta_\delta^B\right| \lesssim \delta^{1-d}$ *and such that*

$$\limsup_{\rho\to+\infty} \rho^{-\frac{d}{d+1}+2+\varepsilon} \|\widehat{\chi}_B(\rho\cdot)\|_{L^1(\Sigma_1)} > 0$$

for any $\varepsilon > 0$.

The proof will be given in the last section.

Remark 6.15. Again, the cases $d = 0$ and $d = 1$ are covered by (6.3) and (6.2) respectively.

Remark 6.16. We point out that the infimum of the numbers d such that $\left|\Delta_\delta^B\right| \lesssim \delta^{1-d}$ is just the upper Minkowski dimension of Δ^B. That is the number

$$d^* = \limsup_{\delta\to 0} \left(\log_{\frac{1}{\delta}} \left(\frac{\left|\Delta_\delta^B\right|}{\delta}\right)\right).$$

It is therefore possible to restate Theorem 6.12 in a form like "Assume $d > d^*$, then (6.14) holds". However we prefer to keep the original statement in Theorem 6.12 for the following two reasons. First, the left-hand side in (6.13) is the quantity which actually arises in the proof. Second, we do not want to confuse naturally different objects, such as the polygons with finitely many sides and certain polygons with infinitely many sides (*e.g.*, with an exponentially decreasing sequence of slopes) which share $d^* = 0$ with the polygons with finitely many sides. For similar reasons we did not introduce the infimum α^* of the α's in Theorem 6.7. On the contrary, we shall introduce α^* and d^* in the following section in order to get a more neat comparison.

6.2.4 Comparing the previous arguments

For any B we denote by d^* the Minkowski dimension of Δ^B (see the above remark). We also denote by α^* the infimum of the α' such that $M_{\rho^{-1}}^B \leq c_\alpha \rho^\alpha$. We have the following theorem.

Theorem 6.17. *Let B be a convex planar body. Then*

$$\alpha^* \leq \frac{d^*}{d^*+1}.$$

Moreover there exists B for which the equality sign holds

The proof will be given in the last section.

Remark 6.18. Theorem 6.17 exhibits an upper bound for α^* in terms of d^*. A lower bound in terms of d^* does not exist in general, since we can construct a family of convex bodies with the same $d^* > 0$ but α^* arbitrarily close to 0.

The proof will be given in the last section.

The situation is different if we add geometric assumptions on B.

Theorem 6.19. *Suppose B is inscribed in a disc* (i.e., *B is the convex hull of a subset of a circle*). *Then $\alpha^* = \frac{d^*}{2}$.*

The proof will be given in the last section.

The circle in the previous statement can be replaced by a closed convex smooth curve with everywhere positive Gaussian curvature.

Remark 6.20. By appealing to Theorem 6.7 and Theorem 6.17 we immediately get the following inequality, which is slightly weaker than the one in Theorem 6.12:

$$\|\widehat{\chi}_B(\rho\cdot)\|_{L^1(\Sigma_1)} \lesssim \rho^{\frac{d}{d+1}-2+\varepsilon}.$$

6.2.5 A lower bound for all convex bodies

The main results in this chapter deal with "intermediate" cases between polygons and convex bodies having a smooth convex arc in the boundary. These cases turn out be extreme. Indeed Podkorytov's theorem is a uniform (with respect to the choice of B) upper bound, while the following theorem gives a uniform lower bound for the L^1-average decay of the Fourier transform.

Theorem 6.21. *Let B be a convex body in $\mathbb{R}^2$, then*

$$\limsup_{\rho\to+\infty} \rho^2 \log^{-1}\rho \, \|\widehat{\chi}_B(\rho\cdot)\|_{L^1(\Sigma_1)} > 0\,.$$

The proof will be given in the last section.

6.3 Applications

Before proceeding to the proofs we shall discuss some applications of the formulated results.

6.3.1 Lattice points

Let B be a planar convex body, let $\sigma \in SO(2)$, and $t \in \mathbb{T}^2$. We consider the discrepancy

$$D_\rho(B) = \mathrm{card}\left(\rho B \cap \mathbb{Z}^2\right) - \rho^2 |B| \tag{6.22}$$

where $|\cdot|$ denotes the area. The results in the previous section and some arguments in [160], [187], [40], and [41] allow us to obtain several upper and lower bounds for averages of the discrepancy (6.22) over rotations or rotations and translations. As a first example, it has been proved in [114], [187], and [40] that, for a polygon P, (6.3) implies

$$\int_{SO(2)} \left|D_\rho(\sigma^{-1}(P))\right| \, d\sigma \lesssim \log^2 \rho.$$

As another example, one can use (6.2) to show that, for any convex planar body B,

$$\left\{\int_{\mathbb{T}^2} \int_{SO(2)} \left|D_\rho(\sigma^{-1}(P) - t)\right|^2 \, d\sigma dt\right\}^{\frac{1}{2}} \lesssim \rho^{\frac{1}{2}}. \tag{6.23}$$

(See, *e.g.*, [114] or [41].) Note that (6.23) is false without the integration in t, as the case of a disc and Hardy's Ω-result (see [116]) show.

Again we focus on the case $p = 1$ and we have the following result, which follows easily from Theorem 6.7 and some known arguments (see, *e.g.*, [114], [187] or [40]).

Theorem 6.24. *Let B be a planar convex body such that $M^B_{\rho^{-1}} \lesssim \rho^\alpha$, with $0 < \alpha < \frac{1}{2}$. Then*

$$\int_{\mathbb{T}^2} \int_{SO(2)} \left|D_\rho(\sigma^{-1}(B) - t)\right| \, d\sigma dt \lesssim \rho^{\frac{2\alpha}{2\alpha+1}} \log \rho. \tag{6.25}$$

Moreover, for every such α there exists a body B satisfying

$$\limsup_{\rho \to +\infty} \rho^{-\alpha+\varepsilon} \int_{\mathbb{T}^2} \int_{SO(2)} \left|D_\rho(\sigma^{-1}(B) - t)\right| \, d\sigma dt > 0,$$

for any $\varepsilon > 0$.

The proof will be given in the last section.

Remark 6.26. The cases $\alpha = 0$ and $\alpha = \frac{1}{2}$ are known, see, *e.g.*, [40] and [41] respectively.

6.3.2 Irregularities of distribution

Suppose $\mathcal{P} = \{z_j\}_{j=1}^{N}$ is a distribution of N points in the unit square $U = [0,1]^2$ treated as the torus $\mathbb{T}^2$. Let B be a convex body in U with diameter smaller than 1. Assume $\varepsilon \leq 1$, $\sigma \in SO(2)$, $t \in \mathbb{T}^2$. The study of the discrepancy

$$D(\mathcal{P}, \varepsilon, \sigma, t) = \sum_{j=1}^{N} \chi_{\varepsilon\sigma^{-1}B-t}(z_j) - N\,\varepsilon^2\,|B|$$

has a long history (see, *e.g.*, the references in [15] and [141, Ch. 6]). A typical result is the following theorem, due to Beck [14] and Montgomery [148, Ch. 6] (see also [40]).

Theorem 6.27. *Let B be a convex body in $U = [0,1]^2$ with diameter smaller than 1. Then there exists $c > 0$, such that for every distribution $\mathcal{P} = \{z_j\}_{j=1}^{N}$ in U.*

$$\left\{\int_0^1 \int_{SO(2)} \int_{\mathbb{T}^2} |D(\mathcal{P}, \varepsilon, \sigma, t)|^2 \, dt \, d\sigma \, d\varepsilon\right\}^{\frac{1}{2}} \gtrsim N^{\frac{1}{4}}.$$

The above result is sharp since Beck and Chen [16] proved the following upper bound.

Theorem 6.28. *Let B be a convex body in $U = [0,1]^2$ with diameter smaller than 1. Then there exists $c > 0$ such that for every positive integer N there exists a distribution $\mathcal{P}$ of N points such that*

$$\left\{\int_0^1 \int_{SO(2)} \int_{\mathbb{T}^2} |D(\mathcal{P}, \varepsilon, \sigma, t)|^2 \, dt \, d\sigma \, d\varepsilon\right\}^{\frac{1}{2}} \lesssim N^{\frac{1}{4}}. \tag{6.29}$$

The above upper bound can be improved after replacing the L^2 norm with the L^1 norm. Indeed, Beck and Chen [17] proved the following result.

Theorem 6.30. *Let P be a convex polygon in $U = [0,1]^2$ with diameter smaller than 1. Then there exists $c > 0$ such that for every positive integer N there exists a distribution $\mathcal{P}$ of N points such that*

$$\int_0^1 \int_{SO(2)} \int_{\mathbb{T}^2} |D(\mathcal{P}, \varepsilon, \sigma, t)| \, dt \, d\sigma \, d\varepsilon \lesssim \log^2 N. \tag{6.31}$$

The following result follows easily from Theorem 6.24, [40] and [41]. The case $\alpha = 0$ provides a different proof of (6.31). In the same way one can get a different proof of the L^2 result in (6.29) too. We point out that appealing to lattice point results does not work for L^p norms when $p > 2$ and the body is a polygon (see [52]).

Theorem 6.32. *Let B be a convex body in $U = [0,1]^2$ with diameter smaller than 1 and such that $M^B_{\rho^{-1}} \lesssim \rho^\alpha$. Then for every positive integer N there exists a distribution $\mathcal{P}$ of N points satisfying*

$$\int_{\mathbb{T}^2} \int_{SO(2)} |D(\mathcal{P},\sigma,t)|\, d\sigma dt \lesssim \begin{cases} \log^2 N & \text{when } \alpha = 0 \\ N^{\frac{\alpha}{1+2\alpha}} \log N & \text{when } 0 < \alpha < \frac{1}{2} \\ N^{\frac{1}{4}} & \text{when } \alpha = \frac{1}{2} \end{cases}$$

where $D(\mathcal{P},\sigma,t) = D(\mathcal{P},1,\sigma,t)$.

The proof will be given in the last section.

6.4 Proofs

The following known result (see, *e.g.*, [46], [155], [41]) will be used throughout.

Lemma 6.33. *Let B be a convex body in $\mathbb{R}^2$. Following the notation in (6.5), we have*

$$|\widehat{\chi}_B(\rho\Theta)| \lesssim \rho^{-1} \left[\left|r(B,\rho^{-1},\theta)\right| + \left|r(B,\rho^{-1},\theta+\pi)\right|\right] ,$$

where $|\cdot|$ denotes the length of the chord.

We define

$$\widetilde{d}(\theta,\Delta^B) = \min\left(d(\theta,\Delta^B), d(\theta+\pi,\Delta^B)\right)$$

and we deduce the following lemma.

Lemma 6.34. *For every $\theta \notin \Delta^B$ we have*

$$|\widehat{\chi}_B(\rho\Theta)| \lesssim \frac{1}{\rho^2 \widetilde{d}(\theta,\Delta^B)} .$$

Proof. Let $\theta \notin \Delta^B$ (say $\theta = -\frac{\pi}{2}$). Assume that ∂B passes through the origin and B lies in the upper half-plane. It follows that in a neighborhood of the origin ∂B is the graph of a non-negative convex function, say $y = \varphi(x)$, satisfying $\varphi(0) = 0$ and $\varphi'(0-) < 0 < \varphi'(0+)$, where $\varphi'(0-)$ and $\varphi'(0+)$ denote the left and the right derivative at the origin respectively. Let

$$E = \left\{(x,y) \in \mathbb{R}^2 : y > \varphi'(0-)x \quad \text{and} \quad y > \varphi'(0+)x\right\}.$$

By convexity $B \subset E$ and therefore

$$\left|r(B,\rho^{-1},\theta)\right| \leq \frac{1}{\rho\varphi'(0+)} + \frac{1}{\rho\left|\varphi'(0-)\right|} \leq \frac{2}{\rho \min\left(\varphi'(0+), |\varphi'(0-)|\right)} .$$

To complete the proof it is enough to observe that

$$\min\left(\varphi'(0+), |\varphi'(0-)|\right) \approx d(\theta,\Delta^B)$$

and to apply the previous lemma. □

The following lemmas will be needed in the proof of Theorem 6.7.

Lemma 6.35. *Let $R \geq 1$, $0 < \beta < \frac{\pi}{4}$. Assume $R\beta < \frac{1}{2}$. Denote by $C = C(\beta, R)$ the convex hull of the set*

$$\{R \exp(i\theta) : -\beta \leq \theta < \beta\} \cup \{P\},$$

where the point P has distance 1 from the points $Re^{\pm i\beta}$ and satisfies $|P| \leq R$. Then there exist positive constants c_1 and c_2 such that if $R\rho\beta^2 \geq c_1$ then we have

$$|\widehat{\chi}_C(\rho\Theta)| \geq c_2 R^{\frac{1}{2}} \rho^{-\frac{3}{2}}$$

for every $|\theta| \leq \frac{\beta}{2}$.

Proof. Integrating by parts, we reduce to estimating

$$\rho^{-1} \int_{\partial C} n(x) \cdot \Theta \exp(2\pi i \rho \Theta \cdot x)\, dx. \tag{6.36}$$

The boundary ∂C consists of two segments and an arc. In order to control the latter we reduce to the oscillatory integral

$$\left| \int_{-R\beta}^{R\beta} \exp\left(i\rho \frac{t^2}{R}\right) dt \right| = \left| R\beta \int_{-1}^{1} \exp\left(i\rho R\beta^2 u\right) du \right| \geq c\, R^{\frac{1}{2}} \rho^{-\frac{1}{2}}$$

for $\rho R \beta^2$ large enough. The two segments have length 1 and their contribution in (6.36) is $O(\rho^{-2})$. □

Lemma 6.37. *Let $R > 1$ and $0 < \beta < \frac{\pi}{4}$. Assume $R\beta < \frac{1}{2}$. For any $N \geq 1$ let $B = B(\beta, R, N)$ be the convex hull of the set*

$$\left\{ R \exp\left(\frac{2\pi i k \beta}{N}\right), k = -N, \dots, N \right\} \cup \{P\}$$

where, as before, the point P has distance 1 from the points $Re^{\pm i\beta}$ and satisfies $|P| \leq R$. Then there exist absolute constants c_1, c_2, and c_3 such that whenever $\rho \geq 2$ and

$$\frac{c_1}{\beta^2} \leq R\rho \leq \frac{c_2}{\beta^2} \frac{N^2}{\log^2 N} \tag{6.38}$$

we have, for any $-\frac{\beta}{2} \leq \theta \leq \frac{\beta}{2}$,

$$|\widehat{\chi}_B(\rho\Theta)| \geq c_3 R^{\frac{1}{2}} \rho^{-\frac{3}{2}}.$$

Proof. Let $C = C(\beta, R)$ be as in Lemma 6.35. By (6.38) and Lemma 6.35 we have

$$|\widehat{\chi}_C(\rho\Theta)| \geq c\, R^{\frac{1}{2}} \rho^{-\frac{3}{2}}$$

when $-\frac{\beta}{2} \leq \theta \leq \frac{\beta}{2}$.

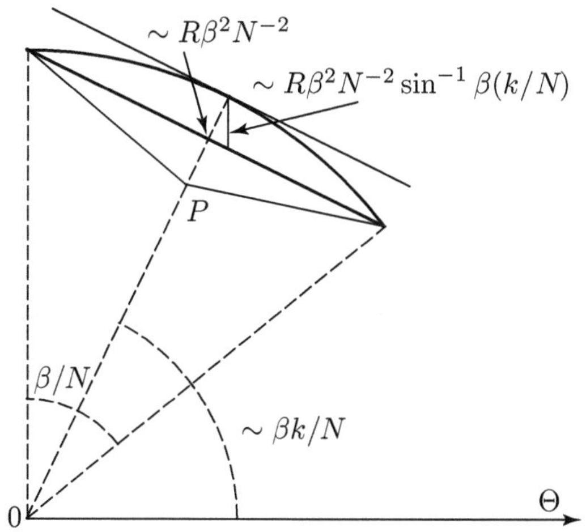

Figure 6.1.

We now study the Fourier transform $\widehat{\chi}_{C\backslash B}$. We claim that

$$\left|\widehat{\chi}_{C\backslash B}(\rho\Theta)\right| \le c\beta\rho^{-1}\frac{\log N}{N}R \tag{6.39}$$

uniformly in θ. Indeed $C\backslash B$ is the union of $2N$ "lunes" $\ell_1, \dots, \ell_{2N}$ (each lune is a convex set bounded by a segment in B and by a portion of the arc in C, see Figure 6.1) and, for any θ,

$$\widehat{\chi}_{C\backslash B}(\rho\Theta) = \widehat{f}(\rho),$$

where $f = f_\theta$ is defined by

$$f(s) = \left|C\backslash B \cap \{\xi \in \mathbb{R}^2 : \xi\cdot\Theta = s\}\right| = \sum_{k=1}^{2N}\left|\ell_k \cap \{\xi \in \mathbb{R}^2 : \xi\cdot\Theta = s\}\right| = \sum_{k=1}^{2N} f_k(s).$$

Note that, for any given s, the above sum contains at most two terms. It is enough to consider one of them, *i.e.*, we assume $0 \le \theta \le \pi$. Moreover we reduce to studying the case $0 \le \theta < \frac{\beta}{N}$, the other cases being similar. In order to bound $\widehat{f}(\rho)$ we estimate the total variation V_f of the function $f(s)$, which is the length of the vertical segment in the kth lune. Now observe that

$$V_{f_k} \le c\beta N^{-1}k^{-1}R$$

whenever $k \ge 1$ (see Figure 6.1).

Summing on k (there are N terms when $\theta = 0$ and $N+1$ terms when $0 < \theta < \frac{\beta}{N}$) we get (6.39).

Finally, for suitable choices of c_1 and c_2 in (6.38) we get

$$\begin{aligned} |\widehat{\chi}_B(\rho\Theta)| &\geq |\widehat{\chi}_C(\rho\Theta)| - |\widehat{\chi}_{B\setminus C}(\rho\Theta)| \\ &\geq c_3 R^{\frac{1}{2}} \rho^{-\frac{3}{2}} - c_4 \beta \rho^{-1} \frac{\log N}{N} R \geq c_5 \rho^{-\frac{3}{2}} R^{\frac{1}{2}}, \end{aligned}$$

as required. □

Proof of Theorem 6.7. We start with the upper bounds in (6.8). Let $P^B_{\rho^{-1}}$ be as in Definition 6.6. Let $\widetilde{P}^B_{\rho^{-1}}$ be the smallest polygon having sides parallel to that of P^B_ρ and containing B. It is not difficult to see that for ρ sufficiently large

$$\left|r(B, \rho^{-1}, \theta)\right| \lesssim \left|r(\widetilde{P}_\rho, c\rho^{-1}, \theta)\right|$$

where again the implicit constant depends only on B. By Lemma 6.33 we have

$$|\widehat{\chi}_B(\rho\Theta)| \lesssim \rho^{-1} \left|r(B, \rho^{-1}, \theta)\right| \lesssim \rho^{-1} \left|r(\widetilde{P}_{\rho^{-1}}, c\rho^{-1}, \theta)\right|.$$

Hence, by the proof of (6.9) in [40] or [41],

$$\rho^{-1} \int_0^{2\pi} \left|r(\widetilde{P}_{\rho^{-1}}, c\rho^{-1}, \theta)\right| d\theta \leq c M^B_{\rho^{-1}} \rho^{-2} \log(\rho) \leq c\rho^{-2+\alpha} \log(\rho)$$

thereby proving (6.8).

We now show that (6.8) is essentially sharp. Let $B = B(\beta, R, N)$ be as in Lemma 6.37 and consider the sets $B_h = B(\beta_h, R_h, N_h)$, $h = 1, 2, 3, \ldots$, where, for any small $\varepsilon > 0$,

$$R_h = 2^{(1-2\alpha)h}, \qquad \beta_h = 2^{h(2\alpha-1-\varepsilon)}, \qquad N_h = 2^{h\alpha}.$$

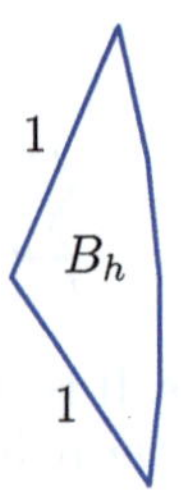

Figure 6.2.

We denote by γ_h the union of the N_h sides and by ζ_h the arc where they are inscribed. Observe that

$$\sum_{h=n_0}^{+\infty} \beta_h R_h < \frac{\pi}{4} \tag{6.40}$$

for a suitable n_0.

We recall that each B_h has the shape in Figure 6.2, *i.e.*, it is a convex polygon consisting of two sides of length 1 and of N_h sides coming from a regular polygon of large radius R_h. Let E_h be the rotated and translated copy of every B_h so that, moving counterclockwise, $E_{n_0} = B_{n_0}$ and two consecutive E_h's have disjoint interior and share a side (of length 1), while the union of the arcs ζ_h's is a convex piecewise smooth curve. We write

$$B = \left(\bigcup_{j=n_0}^{h-1} E_j\right) \cup E_h \cup \left(\bigcup_{j=h+1}^{\infty} E_j\right) = \widetilde{E}_h \cup E_h \cup E_h^{\#}. \tag{6.41}$$

By the condition (6.40) B is a convex set. Let now $\rho_h = 2^h$. Let $p_h = \sum_{j=n_0}^{h} \beta_j$. Being (6.38) satisfied, Lemma 6.35 implies

$$|\widehat{\chi}_{D_h}(\rho_h\Theta)| \geq cR_h^{\frac{1}{2}}\rho_h^{-\frac{3}{2}} = c2^{-h(\alpha+1)}$$

for

$$p_h + \frac{1}{3}\beta_h < \theta < p_h + \frac{2}{3}\beta_h. \tag{6.42}$$

We then estimate the contribution of the convex sets $\widetilde{E}_h$ and $E_h^{\#}$ using Lemma 6.34. Indeed, since θ satisfies (6.42) we obtain, for any h,

$$\left|\widehat{\chi}_{\widetilde{E}_h}(\rho_h\Theta)\right| + \left|\widehat{\chi}_{E_h^{\#}}(\rho_h\Theta)\right| \leq c\beta_h^{-1}\rho_h^{-2}.$$

We then have

$$\begin{aligned}\int_0^{2\pi} |\widehat{\chi}_B(\rho_h\Theta)|\, d\theta &\geq \int_{p_h+\frac{1}{3}\beta_h}^{p_h+\frac{2}{3}\beta_h} |\widehat{\chi}_B(\rho_h\Theta)|\, d\theta \geq \left|c_1\beta_h R_h^{1/2}\rho_h^{-3/2} - c_2\rho_h^{-2}\right| \\ &\geq \left|c_1 2^{h(\alpha-\varepsilon-2)} - c_2 2^{-2h}\right| \geq c_3\rho_h^{-2+\alpha-\varepsilon}.\end{aligned}$$

To complete the proof we estimate $M_{\rho^{-1}}^B$. Given $\rho \geq 2$, let H satisfy $2^H \leq \rho < 2^{H+1}$. Here we split

$$B = \left(\bigcup_{j=n_0}^{H} E_j\right) \cup \left(\bigcup_{j=H+1}^{+\infty} E_j\right) = B_a \cup B_b. \tag{6.43}$$

Observe that the first term is a polygon with $\sum_{j=n_0}^{H} N_j \lesssim 2^{H\alpha}$ sides. Now consider that for any convex polygon Q and any δ the number M_δ^Q cannot exceed the number of sides of Q. Therefore the contribution of B_a to $M_{\rho^{-1}}^B$ is $\lesssim 2^{H\alpha} = \rho^\alpha$. As for B_b we note that the length of $\cup_{j=H+1}^{+\infty}\zeta_j$ is comparable to the length of ζ_H, while the chords of height ρ^{-1} are longer, since $\cup_{j=H+1}^{+\infty}\zeta_j$ comes from flatter arcs. Therefore there are fewer chords than for ζ_H. We have therefore proved that $M_{\rho^{-1}}^B \lesssim \rho^\alpha$. □

Proof of Theorem 6.12. Let $\Omega_\rho = \Delta^B_{\rho^{-\frac{1}{d+1}}}$. In order to estimate

$$I(\rho) = \int_0^{2\pi} |\widehat{\chi}_B(\rho\Theta)|\, d\theta$$

we write

$$I(\rho) = \int_{\Omega_\rho} |\widehat{\chi}_B(\rho\Theta)|\, d\theta + \int_{[0,2\pi]\setminus\Omega_\rho} |\widehat{\chi}_B(\rho\Theta)|\, d\theta = I_1 + I_2.$$

To estimate I_1 we use the Bunyakovskii–Cauchy–Schwarz inequality, the fact that $\left|\Delta_\delta^B\right| \lesssim \delta^{1-d}$, and (6.2):

$$I_1 \leq |\Omega_\rho|^{\frac{1}{2}} \left\{ \int_0^{2\pi} \left|\widehat{\chi}_B(\rho\Theta)\right|^2 d\theta \right\}^{\frac{1}{2}} \lesssim \rho^{\frac{d-1}{2d+2}} \rho^{-\frac{3}{2}} = c\rho^{-2+\frac{d}{d+1}}.$$

In order to estimate I_2 we use Lemma 6.34:

$$\begin{aligned} I_2 &\lesssim \sum_{k=0}^{(d+1)^{-1}\log\rho} \int_{\Delta^B_{2^{-k}} \setminus \Delta^B_{2^{-k-1}}} \frac{c}{\rho^2 \widetilde{d}(\theta, \Delta^B)} d\theta \lesssim \rho^{-2} \sum_{k=0}^{(d+1)^{-1}\log\rho} 2^k \left|\Delta^B_{2^{-k}}\right| \\ &\lesssim \rho^{-2} \sum_{k=0}^{(d+1)^{-1}\log\rho} 2^k 2^{-k(1-d)} \lesssim \rho^{-2} \sum_{k=0}^{(d+1)^{-1}\log\rho} 2^{kd} = c\rho^{-2+\frac{d}{d+1}}. \end{aligned}$$

In order to give a counterexample we use the body B constructed in the proof of Theorem 6.7. Again we consider the sets $B_h = B(\beta_h, R_h, N_h)$, $h = 1, 2, \ldots,$ where now

$$R_h = 2^{h\frac{1-d}{1+d}}, \qquad \beta_h = 2^{h\left(\frac{d-1}{d+1} - \varepsilon\right)}, \qquad N_h = 2^{h\frac{d}{d+1}},$$

while $\rho_h = 2^h$. Arguing as in the proof of the previous theorem we get, for every h,

$$\rho_h^{2-\frac{d}{1+d}+\varepsilon} \int_0^{2\pi} \left|\widehat{\chi}_B(\rho_h\Theta)\right| d\theta \geq c.$$

To complete the proof it is enough to show that $\left|\Delta_\delta^B\right| \lesssim \delta^{1-d}$. We identify Δ_δ^B with a subset of $\left[0, \frac{\pi}{2}\right]$ and we observe that

$$\Delta_\delta^B \cap \left[\sum_{j \leq H-1} \beta_j, \sum_{j \leq H} \beta_j \right]$$

consists of N_H points at distance $\frac{\beta_H}{N_H}$. Given $\delta > 0$, we choose H so that

$$\frac{\beta_H}{N_H} \leq \delta < \frac{\beta_{H-1}}{N_{H-1}}, \quad \text{hence} \quad \beta_H \leq \left(\frac{\beta_H}{N_H}\right)^{1-d} \approx \delta^{1-d}.$$

We now split $B = B_a \cup B_b$ as in (6.43). The contribution of B_a to $\left|\Delta_\delta^B\right|$ is

$$\delta \sum_{j \leq H} N_j \approx \delta N_H \approx \beta_H \lesssim \delta^{1-d},$$

while the contribution of B_b to $\left|\Delta_\delta^B\right|$ is bounded by

$$\sum_{j > H} \beta_j \lesssim \beta_H \lesssim \delta^{1-d},$$

which completes the proof. □

The following proof follows an argument in [171].

Proof of Theorem 6.17. Let ch_j be a side of $P^B_{\rho^{-1}}$ having endpoints x_j and y_j. Assume that moving counterclockwise along the boundary of B the point x_j comes before y_j. Denote with φ_j the direction of the right normal in x_j and with ψ_j the direction of the left normal in y_j. First observe that

$$|ch_j|\,|\varphi_j - \psi_j| \gtrsim \rho^{-1}, \tag{6.44}$$

which follows by convexity when $|\varphi_j - \psi_j| \geq \frac{\pi}{4}$ and by a trigonometric computation when $|\varphi_j - \psi_j| < \frac{\pi}{4}$. Let $\alpha > \alpha^*$. Summing up and applying Hölder inequality we get

$$\begin{aligned}\rho^{-\alpha} M^B_{\rho^{-1}} &\lesssim \sum\nolimits_j |ch_j|^\alpha\,|\varphi_j - \psi_j|^\alpha \leq \left\{\sum\nolimits_j |ch_j|\right\}^\alpha \left\{\sum\nolimits_j |\varphi_j - \psi_j|^{\frac{\alpha}{1-\alpha}}\right\}^{1-\alpha} \\ &\leq |\partial B|^\alpha \left(\sum\nolimits_j |\varphi_j - \psi_j|^{\frac{\alpha}{1-\alpha}}\right)^{1-\alpha}\end{aligned}$$

where the sum is on the $M^B_{\rho^{-1}}$ sides of the polygon $P_{\rho^{-1}}$. It remains to show that $\sum_j |\varphi_j - \psi_j|^{\frac{\alpha}{1-\alpha}}$ is bounded by a constant independent of $P_{\rho^{-1}}$. Let

$$Z_k = \left\{j : 2^{-k}\pi < |\varphi_j - \psi_j| \leq 2^{1-k}\pi\right\}.$$

Now observe that if $j \in Z_k$ then the interval $(\varphi_j, \psi_j) \subseteq \Delta^B_{2^{-k}\pi}$. Now choose d such that $d^* < d < \frac{\alpha}{1-\alpha}$. Then

$$2^{-k}\pi\,\mathrm{card}(Z_k) \leq \left|\Delta^B_{2^{-k}\pi}\right| \lesssim 2^{-k(1-d)},$$

so that $\mathrm{card}(Z_k) \lesssim 2^{kd}$ and therefore

$$\begin{aligned}\sum_j |\varphi_j - \psi_j|^{\frac{\alpha}{1-\alpha}} &\leq \sum_{k=0}^{+\infty} \sum_{j\in Z_k} |\varphi_j - \psi_j|^{\frac{\alpha}{1-\alpha}} \lesssim \sum_{k=0}^{+\infty} 2^{kd} 2^{-k\frac{\alpha}{1-\alpha}} \\ &= \sum_{k=0}^{+\infty} 2^{-k\left(\frac{\alpha}{1-\alpha} - d\right)} < +\infty.\end{aligned}$$

The sharpness of the inequality $\alpha^* \leq \frac{d^*}{d^*+1}$ follows from the common counterexample in the proof of Theorem 6.7 and Theorem 6.12. □

Proof of Remark 6.18. Let $\gamma > 1$ and $\beta > 0$. For $n \geq 1$ let $x_n = n^{-\beta}$ and $y_n = n^{-\beta\gamma}$. Let B denote the convex hull of the infinite points (x_n, y_n). We claim that the polygon $P_{\rho^{-1}}$ associated to B satisfies

$$M^B_{\rho^{-1}} \lesssim \rho^{\frac{1}{\gamma\beta}}$$

(hence $\alpha^* \leq \frac{1}{\gamma\beta}$). Indeed, choose

$$ch_1 = B \cap \left\{ (x,y) \in \mathbb{R}^2 : y = \frac{1}{\rho} \right\}$$

as the first side of $P_{\rho^{-1}}$. The number of sides of B located on the right of ch_1 is $\approx \rho^{\frac{1}{\gamma\beta}}$ and the claim follows since for any polygon D with finitely many sides and any ρ we have $M^D_{\rho^{-1}} \leq \#\,(\text{sides of } D)$. On the other hand one checks that B satisfies

$$\left|\Delta^B_\delta\right| \lesssim \delta^{1-\frac{1}{\beta(\gamma-1)+1}}$$

and the exponent is best possible, *i.e.*,

$$d^* = \frac{1}{\beta(\gamma-1)+1}.$$

If we now choose $\gamma = 1 + \frac{1}{\beta}$ we get $d^* = \frac{1}{2}$ and α^* arbitrarily small (since β can be large). □

Proof of Theorem 6.19. We show that $\alpha^* = \frac{d^*}{2}$ whenever B is inscribed in a disc, namely when B is the convex hull of a subset of a circle.

Let $P^B_{\rho^{-1}}$ be as in Definition 6.6 and assume $\alpha > \alpha^*$, hence $M^B_{\rho^{-1}} \lesssim \rho^\alpha$. Let $x_1, x_2, \ldots$ be the vertices of $P^B_{\rho^{-1}}$. See Figure 6.3.

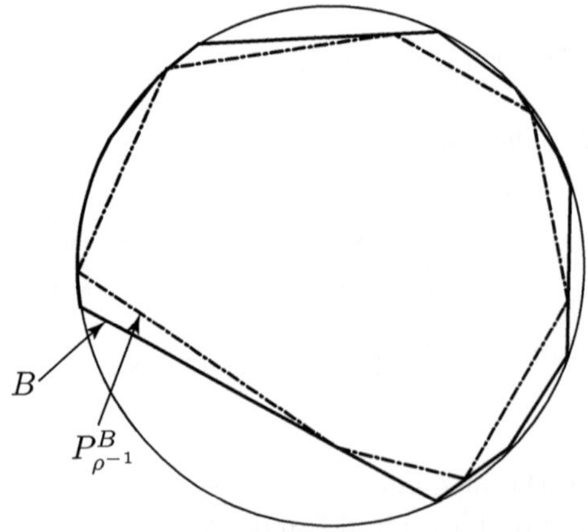

Figure 6.3.

Let $B_1, B_2, \ldots$ be discs of radius $\rho^{-\frac{1}{2}}$ centered at the above vertices. Since B is the convex hull of a subset of a given circle C, there exists a constant c such that, for any j, we are in at least one of the following two cases:

either

i) $cB_j \cup cB_{j+1}$ contains the arc in ∂B connecting x_j and x_{j+1},

or

ii) the part of ∂B connecting x_j and x_{j+1} and not contained in $cB_j \cup cB_{j+1}$ is a segment.

Indeed, assume that i) and ii) fail. Then the arc in ∂B connecting x_j and x_{j+1} must touch the unit circle C outside of the discs cB_j or cB_{j+1}, at a point having distance $\approx \rho^{-1}$ from the side of $P^B_{\rho^{-1}}$ connecting x_j and x_{j+1}. Now observe that this latter can be extended to a chord of C at distance $\approx \rho^{-1}$ from ∂C. Then, for a suitable c, the disc cB_j and cB_{j+1} cannot be distinct.

The above implies that, for $\alpha > \alpha^*$,

$$\Delta^B_{\rho^{-\frac{1}{2}}} \subseteq c_1\pi^{\pm}\left(\partial B \cap \left(\bigcup_{j=1}^{c\rho^\alpha} cB_j\right)\right)$$

and therefore

$$\left|\Delta^B_{\rho^{-\frac{1}{2}}}\right| \lesssim \sum_{j=1}^{c\rho^\alpha} \rho^{-\frac{1}{2}} \approx \rho^{\alpha-\frac{1}{2}} = \left(\rho^{-\frac{1}{2}}\right)^{1-2\alpha},$$

hence, in this case, $d^* \leq 2\alpha^*$.

We now prove that $\alpha^* \leq \frac{d^*}{2}$. Let $\overline{\alpha} < \alpha^*$. Then there exists a sequence $\rho_k \to +\infty$ such that $M^B_{\rho_k^{-1}} \gtrsim \rho_k^{\overline{\alpha}}$. We claim that there exists $\approx \rho_k^{\overline{\alpha}}$ points in Δ^B that are $\approx \rho_k^{-\frac{1}{2}}$ separated. Postponing for a moment the proof of the claim, we conclude that

$$\left|\Delta^B_{\rho_k^{-\frac{1}{2}}}\right| \gtrsim \rho_k^{\overline{\alpha}-\frac{1}{2}} = \left(\rho_k^{-\frac{1}{2}}\right)^{1-2\overline{\alpha}},$$

which implies that the Minkowski dimension d^* of Δ^B cannot be smaller than $2\overline{\alpha}$ and therefore $d^* \geq 2\alpha^*$.

Proof of the claim. Let ch_j, φ_j and ψ_j be as in the proof of Theorem 6.17 and define

$$S_a = \left\{j : |\varphi_j - \psi_j| > \rho_k^{-\frac{1}{2}}\right\}$$
$$S_b = \left\{j : |\varphi_j - \psi_j| \leq \rho_k^{-\frac{1}{2}}\right\}.$$

It is enough to prove that whenever $j \in S_b$ we have

$$|\varphi_j - \psi_j| \gtrsim c\rho_k^{-\frac{1}{2}}.$$

Since B is inscribed in a (unit) circle, a simple geometric argument shows that if

$$|\varphi_j - \psi_j| \leq \rho_k^{-\frac{1}{2}},$$

then the chord ch_j (which is a chord of B of height ρ_k^{-1}) can be continued to a chord of the circle of height $\approx \rho_k^{-1}$ and therefore of length $\approx \rho_k^{-\frac{1}{2}}$. It follows that $|ch_j| \lesssim \rho_k^{-\frac{1}{2}}$ and (6.44) yields

$$|\varphi_j - \psi_j| \gtrsim c\rho_k^{-\frac{1}{2}}$$

for any $j = 1, \dots, c\rho^{\alpha}$. □

The following lemma will be needed in the proof of Theorem 6.21. The proof depends on an easy modification of an argument in [211].

Lemma 6.45. *Let B be a convex planar body containing a large disc of radius r. Let g be a smooth non-negative function supported in the set $\{t+v\}_{t\in B\,,\,|v|\le 1}$ such that $g(t) = 1$ when $t \in B$ and $\operatorname{dist}(t, \partial B) \ge 1$. Then there exists a constant c, independent of r, such that*

$$\|\widehat{g}\|_{L^1(\mathbb{R}^2)} \ge c \log^2 r\,.$$

Proof. We first need the following known inequality (see, *e.g.*, [176] or [56]). Let $h \in L^1(\mathbb{R})$ satisfy $\widehat{h} \in L^1(\mathbb{R})$, $\widehat{h}(u) = 0$ for $u \le 0$. Then

$$\int_{-\infty}^{+\infty} |h(x)|\, dx \ge c \int_1^{+\infty} \frac{1}{u} \left|\widehat{h}(u)\right| du. \tag{6.46}$$

A quick proof of (6.46) follows. Because of [54, p. 584] we can assume $\widehat{h}(u) \ge 0$. We then consider the odd real function s defined by $s(x) = -i\,(1-x)_+$ for $x > 0$, the Fourier transform of which is

$$\widehat{s}(u) = \frac{2\pi u - \sin 2\pi u}{2\pi^2 u^2}.$$

Then

$$\begin{aligned}\int_{-\infty}^{+\infty} |h(x)|\, dx &\ge \left|\int_{-\infty}^{+\infty} h(x)s(x)dx\right| = \left|\int_{-\infty}^{+\infty} \widehat{h}(u)\widehat{s}(u)du\right| \\ &\ge c\int_1^{+\infty} \frac{\widehat{h}(u)}{u} du.\end{aligned}$$

Observe that, through a translation, (6.46) implies the following fact. Suppose $\widehat{h}(u) = 1$ for u in an interval of length r, say $[q, q+r]$. Moreover $\widehat{h}(u) = 0$ for $u \le q-1$, then

$$\int_{-\infty}^{+\infty} |h(x)|\, dx \ge c \log r. \tag{6.47}$$

To prove the lemma we may suppose that B lies in the half-plane $\{(x, y) : x \ge 1\}$ as in Figure 6.4.

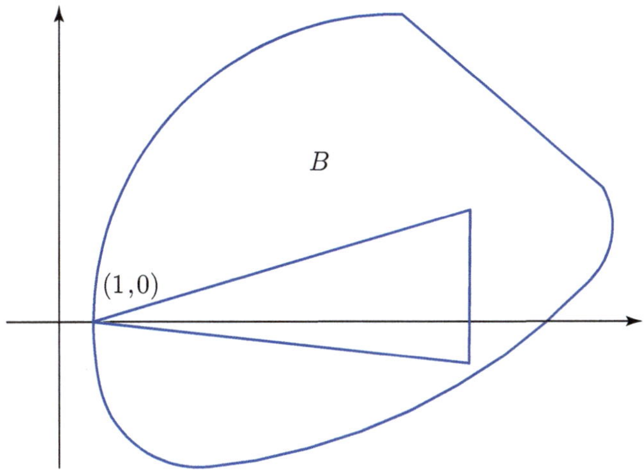

Figure 6.4.

Then, by (6.46) and (6.47),

$$\begin{aligned}\int_{\mathbb{R}}\int_{\mathbb{R}} |\widehat{g}(\xi,\eta)|\, d\xi d\eta &= \int_{\mathbb{R}}\int_{\mathbb{R}} \left|\int_{\mathbb{R}} \left\{\int_{\mathbb{R}} g(x,y)e^{-2\pi i\eta y}dy\right\} e^{-2\pi i\xi x}\, dx\right|\, d\xi d\eta \\ &\geq c\int_{\mathbb{R}}\int_1^{+\infty} \frac{1}{x}\left|\int_{\mathbb{R}} g(x,y)e^{-2\pi i\eta y}dy\right|\, dx d\eta \\ &\geq c\int_1^r \frac{1}{x}\int_{\mathbb{R}}\left|\int_{\mathbb{R}} g(x,y)e^{-2\pi i\eta y}dy\right|\, d\eta dx \geq c\int_1^r \frac{1}{x}\log x\, dx \\ &= c\log^2 r\end{aligned}$$

since, because of the convexity of B, we can assume that $g(x,y)$ takes value 1 inside a whole triangle such as the one in the previous picture. □

Proof Theorem 6.21. Arguing by contradiction we assume the existence of a positive continuous function $\varepsilon(\rho) \to 0$ (as $\rho \to +\infty$), such that

$$\int_0^{2\pi} |\widehat{\chi}_B(\rho\Theta)|\, d\theta \leq \varepsilon(\rho)\rho^{-2}\log\rho \tag{6.48}$$

for $\rho \geq 2$. Let φ be a non-negative radial cut-off function supported in the unit disc, then the convolution

$$g = \chi_{\rho B} * \varphi$$

satisfies the assumptions in the previous lemma (ρB contains a disc of radius $\approx \rho$).

Therefore, by (6.48)

$$\begin{aligned}
\log^2\rho &\le c\,\|\widehat{g}\|_{L^1(\mathbb{R}^2)} = c\rho^2\int_{\mathbb{R}^2}|\widehat{\chi}_B(\rho x)\widehat{\varphi}(x)|\,dx\\
&\le c\rho^2\int_{\mathbb{R}^2}|\widehat{\chi}_B(\rho x)|\,\frac{1}{1+|x|}dx \le c\rho^2\int_0^{+\infty}\frac{u}{1+u}\int_0^{2\pi}|\widehat{\chi}_B(\rho u\Theta)|\,d\theta du\\
&= c\int_0^{+\infty}\frac{s}{1+\rho^{-1}s}\int_0^{2\pi}|\widehat{\chi}_B(s\Theta)|\,d\theta ds \le c\left(1+\int_2^{+\infty}\frac{\varepsilon(s)\log s}{s\,(1+\rho^{-1}s)}ds\right)\\
&\le c\left(1+\int_2^{\rho}\frac{\varepsilon(s)\log s}{s}ds+\rho\int_{\rho}^{+\infty}\frac{\varepsilon(s)\log s}{s^2}ds\right) = A(\rho).
\end{aligned}$$

To end the proof we observe that

$$A(\rho)/\log^2\rho \to 0$$

as $\rho\to+\infty$, by l'Hôspital's rule. □

Remark 6.49. Using an induction argument as in [211], the above theorem can be extended to several variables so that, for any convex body in $\mathbb{R}^n$,

$$\limsup_{\rho\to+\infty}\frac{\rho^n}{\log^{n-1}\rho}\int_{\Sigma_{n-1}}|\widehat{\chi}_B(\rho\sigma)|\,d\sigma > 0\,.$$

Remark 6.50. To prove our theorem we have used an idea introduced in [211] to get lower bounds for Lebesgue constants. Therefore our result shows a relation between the study of Lebesgue constants and the L^1 spherical averages of Fourier transforms of characteristic functions. However we see no general theorem relating one to the other. See [152] for a related discussion with a number theoretic flavor.

Remark 6.51. The estimates of $|r(B,\delta,\theta)|$ (see (6.5)) is a geometrical problem which does not involve necessarily the Fourier transform. The previous theorem and the inequality in Lemma 6.33 imply that, for any convex planar body we have

$$\limsup_{\delta\to0^+}\frac{1}{\delta\log\frac{1}{\delta}}\int_0^{2\pi}|r(B,\delta,\theta)|\,d\theta > 0.$$

The problem considered in the previous remark could be related to the study of floating bodies (see, *e.g.*, [170]), where, in place of fixing δ, one fixes the area ($\approx \delta\,|r(B,\delta,\theta)|$) of the small part of B cut away by the chord $r(B,\delta,\theta)$ in the direction Θ.

Proof of Theorem 6.24. Arguing as in [114] or [40] and applying Theorem 6.7 and (6.2) we have

$$\begin{aligned}
&\int_{\mathbb{T}^2}\int_{SO(2)}\left|D_\rho(\sigma^{-1}(B)-t)\right|\,d\sigma dt\\
&\quad= \rho^2\int_{\mathbb{T}^2}\int_{SO(2)}\left|\sum_{m\neq0}\widehat{\chi}_B(\rho\sigma m)e^{2\pi i m\cdot t}\right|\,d\sigma dt
\end{aligned}$$

$$
\begin{aligned}
&\leq \rho^2 \int_{\mathbb{T}^2} \int_{SO(2)} \left| \sum_{0 \neq |m| \leq \rho^{\frac{1-2\alpha}{1+2\alpha}}} \widehat{\chi}_B(\rho\sigma m) e^{2\pi i m \cdot t} \right| d\sigma dt \\
&\quad + \rho^2 \int_{\mathbb{T}^2} \int_{SO(2)} \left| \sum_{|m| > \rho^{\frac{1-2\alpha}{1+2\alpha}}} \widehat{\chi}_B(\rho\sigma m) e^{2\pi i m \cdot t} \right| d\sigma dt \\
&\leq \rho^2 \sum_{0 \neq |m| \leq \rho^{\frac{1-2\alpha}{1+2\alpha}}} \int_{SO(2)} |\widehat{\chi}_B(\rho\sigma m)| \, d\sigma \\
&\quad + \rho^2 \left\{ \int_{SO(2)} \sum_{|m| > \rho^{\frac{1-2\alpha}{1+2\alpha}}} |\widehat{\chi}_B(\rho\sigma m)|^2 \, d\sigma \right\}^{\frac{1}{2}} \\
&\lesssim \rho^2 \sum_{0 \neq |m| \leq \rho^{\frac{1-2\alpha}{1+2\alpha}}} |\rho m|^{-2+\alpha} \log|\rho m| + \rho^2 \left\{ \sum_{|m| > \rho^{\frac{1-2\alpha}{1+2\alpha}}} |\rho m|^{-3} \right\}^{\frac{1}{2}} \\
&\lesssim \rho^\alpha \int_1^{\rho^{\frac{1-2\alpha}{1+2\alpha}}} t^{\alpha-1} \log(\rho t) dt + \rho^{\frac{1}{2}} \left\{ \int_{\rho^{\frac{1-2\alpha}{1+2\alpha}}}^{+\infty} t^{-2} \right\}^{\frac{1}{2}} \\
&\lesssim \rho^{\frac{2\alpha}{1+2\alpha}}.
\end{aligned}
$$

The lower bound follows from Theorem 6.7 and the orthogonality argument in [40, p. 269]. □

Proof of Theorem 6.32. We prove only the case $0 < \alpha < \frac{1}{2}$. Write N as a sum of four squares: $N = j^2 + k^2 + \ell^2 + m^2$ and let $a_1, a_2, a_3, a_4 \in [0,1)$ be pairwise linearly independent on $\mathbb{Z}$, so that, *e.g.*,

$$a_1 + \frac{p}{j} \neq a_2 + \frac{q}{k}$$

for any choice of the integers p, q, j, k ($j, k \neq 0$). That is

$$\left(a_1 + j^{-1}\mathbb{Z}\right) \cap \left(a_2 + k^{-1}\mathbb{Z}\right) = \emptyset \tag{6.52}$$

when $j \neq k$. Let

$$A_{j^2} = \left\{ \left(a_1 + \frac{p}{j}, \frac{q}{j} \right) \right\}_{p,q \in \mathbb{Z}} \cap \mathbb{T}^2$$

and let us define $A_{k^2}, A_{\ell^2}, A_{m^2}$ accordingly. Define

$$\mathcal{P} = A_{j^2} \cup A_{k^2} \cup A_{\ell^2} \cup A_{m^2}.$$

By (6.52) $\mathcal{P}$ has cardinality N. Since

$$\begin{aligned}&\operatorname{card}(\mathcal{P}\cap B)-N\,|B|\\&\quad=\operatorname{card}\left(A_{j^2}\cap B\right)-j^2\,|B|+\cdots+\operatorname{card}\left(A_{m^2}\cap B\right)-m^2\,|B|,\end{aligned}$$

it is enough to prove that, say,

$$\int_{\mathbb{T}^2}\int_{SO(2)}\left|\operatorname{card}\left(A_{j^2}\cap(\sigma(B)+t)\right)-j^2\,|B|\right|\,d\theta dt\lesssim N^{\frac{\alpha}{1+2\alpha}}\log N.$$

We can therefore prove the theorem assuming N to be a square, say $N=r^2$, $r\in\mathbb{N}$ and

$$\mathcal{P}=A_N=\left\{\left(a+\frac{p}{r},\frac{q}{r}\right)\right\}_{p,q\in\mathbb{Z}^2}\cap U.$$

Now observe that, writing $w=(a,0)$ and applying Theorem 6.24, we have

$$\begin{aligned}&\int_{\mathbb{T}^2}\int_{SO(2)}|D(\mathcal{P},\theta,t)|\,dt\,d\sigma\\&\quad=\int_{SO(2)}\int_{\mathbb{T}^2}\left|\operatorname{card}\left(A_{r^2}\cap(\sigma(B)+t)\right)-r^2\,|B|\right|\,dt\,d\sigma\\&\quad=\int_{SO(2)}\int_{\mathbb{T}^2}\left|\operatorname{card}\left(A_{r^2}\cap(\sigma(B)+t+w)\right)-r^2\,|B|\right|\,dt\,d\sigma\\&\quad=\int_{SO(2)}\int_{\mathbb{T}^2}\left|\operatorname{card}\left(\left\{\left(\frac{p}{r},\frac{q}{r}\right)\right\}_{p,q=0}^{r-1}\cap(\sigma(B)+u)\right)-r^2\,|B|\right|\,du\,d\sigma\\&\quad=\int_{SO(2)}\int_{\mathbb{T}^2}\left|\operatorname{card}\left(\mathbb{Z}^2\cap(r\sigma(B)+ru)\right)-r^2\,|B|\right|\,du\,d\sigma\\&\quad=\int_{SO(2)}\int_{\mathbb{T}^2}\left|\operatorname{card}\left(\mathbb{Z}^2\cap(r\sigma(B)+u)\right)-r^2\,|B|\right|\,du\,d\sigma\\&\quad\lesssim r^{\frac{2\alpha}{1+2\alpha}}\log r\\&\quad=\frac{1}{2}N^{\frac{\alpha}{1+2\alpha}}\log N,\end{aligned}$$

where we have used the fact that for a function $f\in L^1\left(\mathbb{T}^2\right)$ and for any integer $k\neq 0$,

$$\int_{\mathbb{T}^2}f(ku)du=\int_{\mathbb{T}^2}f(u)du,$$

which completes the proof. □

The above argument extends to several variables after replacing the sum of four squares with Hilbert's theorem (Waring's problem).

Chapter 7

Geometry of the Gauss Map and Lattice Points in Convex Domains

In the previous two chapters, we have gained a significant amount of understanding about the L^p-average decay for the Fourier transform of characteristic functions of convex sets and considered some applications to problems in lattice point counting and discrepancy theory. In this chapter we consider more elaborate applications of average decay in number theory where the discrepancy function needs to be estimated for almost every rotation instead of averaging over rotations in some L^p-norm. This naturally leads us to the examination of certain maximal functions and as a result brings in some classical harmonic analysis that arises so often in the first part of this book.

7.1 Two main results

Let Ω be a convex planar domain, and let $N(R) = \text{card}\{R\Omega \cap \mathbb{Z}^2\}$. It was observed by Gauss that $N(R) = |\Omega|R^2 + D(R)$, where $|D(R)| \lesssim R$, since the discrepancy $D(R)$ cannot be larger than the number of lattice points that live a distance at most $\frac{1}{\sqrt{2}}$ from the boundary of Ω. As above, here, and throughout the chapter, $A \lesssim B$ means that there exists a uniform C, such that $A \leq CB$. Similarly, $A \approx B$ means that $A \lesssim B$ and $B \lesssim A$. For a general domain this estimate cannot be improved, as can be seen by taking Ω to be a square with sides parallel to the axis. However, the purpose of our further consideration is to show that the remainder term is better for almost every rotation of the domain.

If the boundary of Ω has everywhere non-vanishing Gaussian curvature, better estimates for the remainder term are possible. It is a classical result of Hlawka and Herz that, in that case, $|D(R)| \lesssim R^{\frac{2}{3}}$. An example due to Jarnik shows that without further assumptions, this result is best possible. See, for example, [117] and [112]. If the boundary is assumed to have a certain degree of smoothness, further improvements have been obtained, culminating (at the moment) in a result due to Huxley, see [102], which says that if the boundary of Ω is five times differentiable and has curvature bounded below by a fixed constant, then $|D(R)| \lesssim R^{\frac{46}{73}}$. This is also the current best result for the circle problem, for which the well-known conjecture is that $|D(R)| \lesssim R^{\frac{1}{2}+\epsilon}$ and indeed it was observed by Hardy that one cannot do better than $R^{\frac{1}{2}}$ times an appropriate power of the logarithm.

We have noted that in general the trivial estimate, $|D(R)| \lesssim R$ cannot be improved without curvature assumption on the boundary of the domain. For example, it was proved by Randol in [160], that if Ω is given by the equation $x_1^m + x_2^m \leq 1$, $m > 2$, then $|D(R)| \lesssim R^{\frac{m-1}{m}}$, and $\frac{m-1}{m}$ cannot be replaced by any smaller number. On the other hand, Colin de Verdière showed in [52] that if the boundary of Ω has finite order of contact with its tangent lines, then, for almost every rotation of Ω, the corresponding error term $|D(R,\theta)| \lesssim R^{\frac{2}{3}}$. This result was extended to a certain class of domains, where the order of contact is infinite, in [106]. This raises the obvious question of whether this result holds for an arbitrary convex planar domain. Up to a logarithmic transgression, we answer this question in the affirmative. This is the substance of the next result.

Theorem 7.1. *Let Ω be a convex domain, and let $\delta > \frac{1}{2}$. Define*

$$\mathcal{M}(\theta) = \sup_{R \geq 2} \log^{-\delta}(R) R^{-\frac{2}{3}} |D(R,\theta)|,$$

where $D(R,\theta)$ is the discrepancy corresponding to the domain Ω dilated by R and rotated by the angle θ. Then $\mathcal{M}(\theta) < \infty$ for almost every θ. More precisely $\mathcal{M} \in Weak - L^2(S^1)$, i.e.

$$|\{\theta \in S^1 : \mathcal{M}(\theta) > t\}| \lesssim t^{-2}.$$

Theorem 7.1 is begging to be generalized for the following reason. A result due to Skriganov, see [175], says that if Ω is a polygon, $|D(R,\theta)| \lesssim \log^{1+\epsilon}(R)$, for any $\epsilon > 0$, for almost every θ. There is much room between this result and $R^{\frac{2}{3}} \log^{\frac{1}{2}+\epsilon}(R)$ that we obtain above, and it makes one ask which geometric properties are in play here. We address this issue in the following way. At every point of a convex set there is the left and the right tangent. Therefore, at every point we have the left($-$) and the right($+$) normal. Let $\mathcal{N}^{\pm} : \partial\Omega \to S^1$ denote the Gauss maps, which take each point on the boundary of Ω to the right/left unit normal at that point. Our second main result is the following.

Theorem 7.2. *Let Ω be a convex domain. Let $\mathcal{N}^{\pm}$ be the Gauss maps defined above, and let*

$$\mathcal{N}(\partial\Omega) = \mathcal{N}^{+}(\partial\Omega) \cup \mathcal{N}^{-}(\partial\Omega).$$

Suppose there exists $0 \leq d \leq 1$ such that for any sufficiently small ϵ,

$$|\{\theta \in S^1 : \text{dist}(\theta, \mathcal{N}(\partial\Omega)) \leq \epsilon\}| \lesssim \epsilon^{1-d}. \tag{7.3}$$

Define

$$\mathcal{M}_d(\theta) = \sup_{R\geq 2} \log^{-\delta}(R) R^{-\frac{2d}{2d+1}} |D(R,\theta)|.$$

Then $\mathcal{M}_d \in L^1(S^1)$ if $d > 0$ and $\delta > 1$, or, if $d = 0$ and $\delta > 3$. In particular, $\mathcal{M}_d(\theta) < \infty$ for almost every θ.

The estimate (7.3) implies that the upper Minkowski dimension of $\mathcal{N}(\partial\Omega)$ is at most d. Conversely, if the upper Minkowski dimension of $\mathcal{N}(\partial\Omega)$ is d, then for any $\eta > 0$ the estimate holds, up to an arbitrarily small power of ϵ,

$$|\{\theta \in S^1 : \mathrm{dist}(\theta, \mathcal{N}(\partial\Omega)) \leq \epsilon\}| \lesssim \epsilon^{1-d-\eta}. \tag{7.4}$$

The conclusion of Theorem 7.2 can be improved under additional assumptions. For example, see [44], if Ω is the convex hull of a subset of a circle, then one can replace $\frac{d}{d+1}$ in Lemma 7.13 below by $\frac{d}{2}$ and this changes the exponent $\frac{2d}{2d+1}$ in Theorem 7.2 to $\frac{2d}{d+2}$.

Theorem 7.2 is stated in terms of the estimate (7.4) for the sake of simplicity, but it could be restated somewhat more precisely in terms of the properties of the distribution function $|\{\theta \in S^1 : \mathrm{dist}(\theta, \mathcal{N}(\partial\Omega)) \leq \epsilon\}|$. When $d = 0$ the condition (7.4) defines a polygon with finitely many sides but, as we said, in this case a better result is known. The case $d > 0$ includes polygons with infinitely many sides and also more complicated bodies.

7.2 Examples and preliminary results

Let us begin with some easy examples.

7.2.1 Examples

Example 7.5. Our first example illustrates the case $d > 0$ of Theorem 7.2. Consider a polygon with infinitely many sides, where the slopes of the normals to the sides form a sequence $\{j^{-\alpha}\}_{j=1,2,\ldots}$. It is not hard to see that the upper Minkowski dimension of $\mathcal{N}(\partial\Omega)$ is $\frac{1}{1+\alpha}$ and also (7.4) holds with $d = \frac{1}{1+\alpha}$.

Example 7.6. We now consider the case of a polygon with infinitely many sides, such that the slopes of the normals form a lacunary sequence, for example, the sequence $\{2^{-j}\}_{j=0,1,\ldots}$. In this case, the upper Minkowski dimension of $\mathcal{N}(\partial\Omega)$ is 0, whereas the estimate (7.4) does not hold with $d = 0$, though it holds for every $d > 0$. So, Theorem 0.2 says that for every positive ν,

$$\sup_{R\geq 2} R^{-\nu} |D(R,\theta)|$$

is finite for almost every θ. However, as we foreshadowed above, we can do better if we work directly with the quantity

$$|\{\theta \in S^1 : \operatorname{dist}(\theta, \mathcal{N}(\partial\Omega)) \le \epsilon\}|,$$

instead of the condition (7.4). It is not difficult to see that, in this case,

$$|\{\theta \in S^1 : \operatorname{dist}(\theta, \mathcal{N}(\partial\Omega)) \le \epsilon\}| \approx \epsilon \log\left(\frac{1}{\epsilon}\right). \tag{7.7}$$

The estimate (7.7) along with the proof of Theorem 7.2 yields the conclusion of Theorem 7.2 with $d = 0$ and $\delta > 4$.

We note that Theorem 7.2 does not apply only to polygons, with finitely, or infinitely many sides. In fact, it is not difficult to construct examples of convex domains, where $\mathcal{N}(\partial\Omega)$ has upper Minkowski dimension $0 < d < 1$, which are not polygons. It is just a matter of constructing an appropriate increasing function, for example a Cantor–Lebesgue type function, which defines the tangent vector field.

As we have seen the quest for the best exponent in lattice point problems has a long history and it seems far from definitive results. Also the exponent $\frac{2}{3}$ in Theorem 7.1 is probably not sharp and a natural conjecture is $\frac{1}{2} + \epsilon$. This belief is supported by the fact that

$$\left(\int_{S^1}\int_{\mathbb{T}^2} |D(R,\theta,\tau)|^2 d\tau d\theta\right)^{\frac{1}{2}} \approx R^{\frac{1}{2}},$$

where $\mathbb{T}^2$ denotes the two-dimensional torus, and $D(R,\theta,\tau)$ denotes the discrepancy corresponding to the case where a convex domain Ω is rotated by θ, and translated by τ. See, *e.g.*, [160] or [41, Theorem 6.2].

7.2.2 Estimates for the Fourier transform

The main ingredient in the proof of Theorem 7.1 and Theorem 7.2 is the following maximal stationary phase estimate for the Fourier transform of the characteristic function of Ω, which is interesting in its own right. Under the analyticity assumption, this estimate is implied by a result obtained by Svensson in [185]. However, lack of any smoothness assumption, besides convexity, involves considerable difficulties.

Theorem 7.8. *Let Ω be a convex domain. Then*

$$\left|\left\{\theta \in S^1 : \sup_{R \ge 0} R^{\frac{3}{2}} |\widehat{\chi}_\Omega(R\theta)| > t\right\}\right| \lesssim t^{-2}.$$

The above inequality means that the function

$$\sup_{R\geq 0} R^{\frac{3}{2}}|\widehat{\chi}_\Omega(R\theta)|$$

belongs to $Weak - L^2(S^1)$. Note that in general it does not belong to $L^2(S^1)$, as one can check by assuming Ω to be a square.

In order to prove Theorem 7.2 we shall also use the following decay estimate, the proof of which is taken from [44].

Theorem 7.9. *Let Ω be a convex domain. Then*

$$|\widehat{\chi}_\Omega(R\theta)| \lesssim R^{-2}(\text{dist}(\theta, \mathcal{N}(\partial\Omega))^{-1}.$$

7.3 Proofs

Each subsection of this section contains a proof of the corresponding theorem.

7.3.1 Proof of Theorem 7.1

The proof of this theorem as long as the proof of Theorem 7.2 is a consequence of the techniques in [96] and [93] along with the maximal estimate for the Fourier transform in Theorem 7.8.

Let ψ be a smooth positive radial function of mass 1 supported in the unit disc centered at the origin. Let

$$\psi_\epsilon(x) = \epsilon^{-2}\psi(\epsilon^{-1}x).$$

Define

$$N(R,\theta,\epsilon) = \sum_{k\neq(0,0)} \chi_{R\theta^{-1}\Omega} * \psi_\epsilon(k),$$

and let

$$D(R,\theta,\epsilon) = N(R,\theta,\epsilon) - R^2|\Omega|.$$

Lemma 7.10. *We have*

$$D(R,\theta,\epsilon) = R^2 \sum_{k\neq(0,0)} \widehat{\chi}_\Omega(R\theta k)\widehat{\psi}(\epsilon k).$$

Proof. This is the Poisson summation formula (see Ch. 1). □

Lemma 7.11. *We have*

$$D(R-\epsilon,\theta,\epsilon) - (2R\epsilon - \epsilon^2)|\Omega| \leq D(R,\theta) \leq D(R+\epsilon,\theta,\epsilon) - (2R\epsilon + \epsilon^2)|\Omega|.$$

Proof. We may assume that Ω contains the origin. We have

$$\chi_{(R-\epsilon)\theta^{-1}\Omega} * \psi_\epsilon(k) \leq \chi_{R\theta^{-1}\Omega}(k) \leq \chi_{(R+\epsilon)\theta^{-1}\Omega} * \psi_\epsilon(k),$$

along with

$$N(R-\epsilon,\theta,\epsilon) \leq N(R,\theta) \leq N(R+\epsilon,\theta,\epsilon),$$

and the result follows. □

Lemma 7.12. *We have*

$$\left|\left\{\theta \in S^1 : \sup_{2^j \leq R \leq 2^{j+1}} R^{-\frac{2}{3}}|D(R,\theta)| > t\right\}\right| \lesssim t^{-2}.$$

Proof. We have

$$\sup_{2^j \leq R \leq 2^{j+1}} |D(R,\theta,\epsilon)| \leq \epsilon^{\frac{3}{2}} 2^{\frac{j+1}{2}} \sum_{k \neq (0,0)} |\epsilon k|^{-\frac{3}{2}} |\widehat{\psi}(\epsilon k)| \sup_{2^j \leq R \leq 2^{j+1}} |R\theta k|^{\frac{3}{2}} |\widehat{\chi}_\Omega(R\theta k)|.$$

Since the function

$$\sup_{2^j \leq R \leq 2^{j+1}} |R\theta k|^{\frac{3}{2}} |\widehat{\chi}_\Omega(R\theta k)|$$

is uniformly in Weak $-L^2$, by Theorem 7.8, the sum is also in this space, with the norm controlled by

$$\epsilon^{\frac{3}{2}} 2^{\frac{j}{2}} \sum_{k \neq (0,0)} |\epsilon k|^{-\frac{3}{2}} |\widehat{\psi}(\epsilon k)| \lesssim 2^{\frac{j}{2}} \epsilon^{-\frac{1}{2}}.$$

The result now follows from Lemma 7.11 by taking $\epsilon = 2^{-\frac{j}{3}}$. □

We are now ready to complete the proof of Theorem 7.1. Observe that

$$\sup_{R \geq 2} \log^{-\delta}(R) R^{-\frac{2}{3}} |D(R,\theta)| \lesssim \left\{\sum_{j=1}^{\infty} j^{-2\delta} \sup_{2^j \leq R \leq 2^{j+1}} R^{-\frac{4}{3}} |D(R,\theta)|^2\right\}^{\frac{1}{2}}.$$

The function

$$\sup_{2^j \leq R \leq 2^{j+1}} R^{-\frac{4}{3}} |D(R,\theta)|^2$$

is uniformly in Weak $-L^1$, and can therefore be summed by the sequence $j^{-2\delta}$, $2\delta > 1$. The conclusion of Theorem 7.1 follows. □

7.3.2 Proof of Theorem 7.2

Lemma 7.13. *Under the assumptions of Theorem 7.2, we have*

$$\int_{S^1}\left[\sup_{2^j\le R\le 2^{j+1}} R^{2-\frac{d}{d+1}}|\widehat{\chi}_\Omega(R\theta)|\right]d\theta \lesssim 1,$$

if $d>0$, and if $d=0$,

$$\int_{S^1}\left[\sup_{2^j\le R\le 2^{j+1}} R^{2}|\widehat{\chi}_\Omega(R\theta)|\right]d\theta \lesssim j.$$

Proof. Applying Theorem 7.8, we have

$$\begin{aligned}
&\int_{\{d(\theta,\mathcal{N}(\partial\Omega))\le 2^{-\frac{j}{d+1}}\}}\left[\sup_{2^j\le R\le 2^{j+1}} R^{2-\frac{d}{d+1}}|\widehat{\chi}_\Omega(R\theta)|\right]d\theta\\
&\lesssim 2^{j(\frac{1}{2}-\frac{d}{d+1})}\int_{\{d(\theta,\mathcal{N}(\partial\Omega))\le 2^{-\frac{j}{d+1}}\}}\left[\sup_{2^j\le R\le 2^{j+1}} R^{\frac{3}{2}}|\widehat{\chi}_\Omega(R\theta)|\right]d\theta\\
&\lesssim 2^{j(\frac{1}{2}-\frac{d}{d+1})}|\{\theta\in S^1 : d(\theta,\mathcal{N}(\partial\Omega))\le 2^{-\frac{j}{d+1}}\}|^{\frac{1}{2}}\\
&\lesssim 1.
\end{aligned}$$

Moreover, by Theorem 7.9, we have

$$\begin{aligned}
&\int_{\{d(\theta,\mathcal{N}(\partial\Omega))>2^{-\frac{j}{d+1}}\}}\left[\sup_{2^j\le R\le 2^{j+1}} R^{2-\frac{d}{d+1}}|\widehat{\chi}_\Omega(R\theta)|\right]d\theta\\
&\lesssim 2^{-j\frac{d}{d+1}}\int_{\{d(\theta,\mathcal{N}(\partial\Omega))>2^{-\frac{j}{d+1}}\}}(d(\theta,\mathcal{N}(\partial\Omega)))^{-1}d\theta\\
&\lesssim 2^{-j\frac{d}{d+1}}\sum_{h=0}^{\infty}\left(2^{h-\frac{j}{d+1}}\right)^{-1}|\{\theta\in S^1 : 2^{h-\frac{j}{d+1}}\\
&\le d(\theta,\mathcal{N}(\partial\Omega))\le 2^{h+1-\frac{j}{d+1}}|\\
&\lesssim 2^{-j\frac{d}{d+1}}\sum_{h=0}^{\infty}\left(2^{h-\frac{j}{d+1}}\right)^{-1}\left(2^{h-\frac{j}{d+1}}\right)^{1-d}\\
&\lesssim 1.
\end{aligned}$$

Observe that when $d=0$, it suffices to sum the series in the range $0\le h\lesssim j$. This completes the proof of Lemma 7.13. □

Lemma 7.14. *Under the assumptions of Theorem 7.2, we have*

$$\int_{S^1}\left[\sup_{2^j\le R\le 2^{j+1}} R^{-\frac{2d}{2d+1}}|D(R,\theta)|\right]d\theta \lesssim 1,$$

if $d>0$.

If $d = 0$, we have

$$\int_{S^1}\left[\sup_{2^j \le R \le 2^{j+1}} |D(R,\theta)|\right] d\theta \lesssim j^2.$$

Proof. Assume that $d > 0$. We have

$$\int_{S^1}\left[\sup_{2^j \le R \le 2^{j+1}} |D(R,\theta,\epsilon)|\right] d\theta$$

$$\lesssim \epsilon^{2-\frac{d}{d+1}} 2^{\frac{jd}{d+1}} \sum_{k \neq (0,0)} |\epsilon k|^{-2+\frac{d}{d+1}} |\widehat{\psi}(\epsilon k)| \int_{S^1}\left[\sup_{2^j \le R \le 2^{j+1}} |R\theta k|^{2-\frac{d}{d+1}} |\widehat{\chi}_\Omega(R\theta k)|\right] d\theta$$

$$\lesssim \epsilon^{-\frac{d}{d+1}} 2^{j\frac{d}{d+1}}.$$

We also have

$$\int_{S^1}\left[\sup_{2^j \le R \le 2^{j+1}} R^{-\frac{2d}{2d+1}} |D(R,\theta)|\right] d\theta$$

$$\lesssim 2^{-2d\frac{j}{2d+1}} \left(\int_{S^1}\left[\sup_{2^j \le R \le 2^{j+1}} |D(R,\theta,\epsilon)|\right] d\theta + 2^j\epsilon\right)$$

$$\lesssim 2^{-2d\frac{j}{2d+1}} (\epsilon^{-\frac{d}{d+1}} 2^{j\frac{d}{d+1}} + 2^j\epsilon).$$

Choosing $\epsilon = 2^{-\frac{j}{2d+1}}$ yields the required result. The proof in the case $d = 0$ is similar. □

We are now ready to complete the proof of Theorem 7.2. We have

$$\sup_{R \ge 2} \log^{-\delta}(R) R^{-2\frac{d}{2d+1}} |D(R,\theta)| \lesssim \sum_{j=1}^{\infty} j^{-\delta} \sup_{2^j \le R \le 2^{j+1}} R^{-2\frac{d}{2d+1}} |D(R,\theta)|.$$

In view of Lemma 7.14, if $d > 0$, the series converges for $\delta > 1$ and, if $d = 0$, one must take $\delta > 3$. This completes the proof of Theorem 7.2. □

7.3.3 Proof of Theorem 7.8

We start out by arguing that we may take Ω with a smooth boundary and everywhere non-vanishing curvature, so long as the constants in our argument do not depend on curvature and smoothness. Indeed, suppose that

$$\sup_{R>0} R^{\frac{3}{2}} |\widehat{\chi}_\Omega(R\theta)|$$

is not in Weak $-L^2(S^1)$. This means that given $k > 0$, there exists $N > 0$, such that the weak $-L^2(S^1)$ norm of

$$\sup_{0<R<N} R^{\frac{3}{2}} |\widehat{\chi}_\Omega(R\theta)|$$

is at least k. Approximate Ω by a sequence Ω_n of convex domains such that the boundary of each Ω_n is smooth and has everywhere non-vanishing curvature. We arrange things so that $|\Omega - \Omega_n| \le \frac{1}{n}$ and the claim follows by taking n sufficiently large.

We fix a direction θ, and without loss of generality we assume $\theta = (1,0)$. Then

$$\widehat{\chi}_\Omega(R,0) = \int_{-\infty}^{+\infty} \left(\int_{-\infty}^{+\infty} \chi_\Omega(x_1,x_2)dx_2 \right) e^{-2\pi i x_1 R} dx_1 = \widehat{h}(R),$$

where $h(s)$ denotes the length of the segment obtained by intersecting Ω with the line $x_1 = s$. This function is concave on a suitable interval $[a,b]$. Applying [155] or Lemma 3.7 of [41], we obtain

$$\begin{aligned} |\widehat{\chi}_\Omega(R\theta)| &\le \frac{c}{R} \left(h\left(a + \frac{1}{2R}\right) + h\left(b - \frac{1}{2R}\right)\right) \\ &= \frac{c}{R} \left(\mu\left(\theta, \frac{1}{2R}\right) + \mu\left(-\theta, \frac{1}{2R}\right)\right), \end{aligned}$$

where $\mu(\theta,\epsilon)$ denotes the length of the chord

$$C(\theta,\epsilon) = \{x \in \Omega : x \cdot \theta = S_\theta - \epsilon\},$$

and $S_\theta = \sup_{x\in\Omega} x \cdot \theta$. We are therefore reduced to studying the maximal function

$$\mu^*(\theta) = \sup_{\epsilon>0} \frac{1}{\sqrt{\epsilon}} \mu(\theta,\epsilon).$$

Observe $C(\theta,0)$ is a single point, which we denote by $z(\theta)$. Let now θ_o be a fixed direction. With a mild abuse of notation let $\theta = e^{i\theta}$. We denote by $\lambda(\theta)$ the arc-length on $\partial\Omega$ between $z(\theta_o)$ and $z(\theta)$. Let

$$\lambda^*(\theta) = \sup_{\theta\neq\phi} \frac{\lambda(\theta) - \lambda(\phi)}{\theta - \phi}.$$

We shall need the following estimate.

Lemma 7.15. *We have*

$$[\mu^*(\theta)]^2 \le 4\lambda^*(\theta).$$

Proof. The normal at the point $z(\theta)$ determines (on the chord $C(\theta,\epsilon)$, or possibly on its continuation) two segments of length $\mu_1(\theta,\epsilon)$ and $\mu_2(\theta,\epsilon)$ and the maximal function $\mu^*(\theta)$ is dominated by the sum $\mu_1^*(\theta)+\mu_2^*(\theta)$. Now observe that the computation of $\sup_{\epsilon>0}\frac{1}{\sqrt{\epsilon}}\mu_j(\theta,\epsilon)$ involves only values of ϵ for which $\mu_j(\theta,\epsilon)$ increases. Hence, we may assume that the boundary of Ω is locally a graph of a function $f(x)$, with $f(0)=f'(0^+)=0$, $0\le x\le a$ and

$$[\mu_j^*(\theta)]^2 \le \sup_{0<x<a}\frac{x^2}{f(x)}.$$

By the mean value theorem,

$$\begin{aligned}\sup_{0<x<a}\frac{x^2}{f(x)} &\le \sup_{0\le z\le a}\frac{2z}{f'(z)} \le \sup_{0\le z\le a}\frac{2}{f'(z)}\int_0^z\sqrt{1+(f'(t))^2}\,dt\\ &= \sup_{0\le\psi\le f'(a)}\frac{2}{\psi}\int_0^{(f')^{-1}(\psi)}\sqrt{1+(f'(t))^2}\,dt\\ &= 2\sup_{0\le\psi\le f'(a)}\frac{\lambda(\psi+\theta)-\lambda(\theta)}{\psi} = 2\lambda^*(\theta).\end{aligned}$$

This completes the proof of Lemma 3.1. □

Theorem 7.8 now follows from the classical Hardy–Littlewood maximal theorem, which we state in the following form.

Lemma 7.16. *Let λ be and increasing bounded function on the interval $[a,b]$. Then for every $t>0$,*

$$|\{\theta:\lambda^*(\theta)>t\}| \le \frac{\lambda(b)-\lambda(a)}{t}. \tag{7.17}$$

7.3.4 Proof of Theorem 7.9

By Lemma 3.8 in [41], we have

$$|\widehat{\chi}_\Omega(R\theta)| \lesssim \left(|A_\Omega(R^{-1},\theta)| + |A_\Omega(R^{-1},\theta+\pi)|\right),$$

where

$$A_\Omega(R^{-1},\theta) = \{x\in\Omega : S_\theta-\epsilon < x\cdot\theta < S_\theta\}$$

and

$$S_\theta = \sup_{x\in\Omega} x\cdot\theta,$$

as in the proof of Theorem 7.8.

Without loss of generality we may assume that $\theta=-\frac{\pi}{2}$. We may also assume that the boundary of Ω passes through the origin and Ω lies in the upper half-plane. In a neighborhood of the origin the boundary of Ω is described by a convex function, say $y = \phi(x)$, satisfying $\phi(x) \geq 0$, and $\phi(0) = 0$. Let $\phi'(0^-)$ and $\phi'(0^+)$ denote the left and the right derivatives at the origin. Since we assume that $\theta \notin \mathcal{N}(\partial\Omega)$, we must have $\phi'(0^-) < 0 < \phi'(0^+)$. Let

$$C = \{(x, y) \in \mathbb{R}^2 : y > \phi'(0^-)x \text{ and } y > \phi'(0^+)x\}.$$

By convexity, $\Omega \subset C$. It follows that

$$\begin{aligned} |A_\Omega(R^{-1}, \theta)| &\leq \frac{1}{R^2\phi'(0^+)} + \frac{1}{R^2|\phi'(0^-)|} \\ &\leq \frac{2}{R^2 \min\{\phi'(0^+), |\phi'(0^-)|\}}. \end{aligned}$$

Since

$$\min\{\phi'(0^+), |\phi'(0^-)|\} \approx \operatorname{dist}(\theta, \mathcal{N}(\partial\Omega)),$$

the proof follows. □

Chapter 8

Average Decay Estimates for Fourier Transforms of Measures Supported on Curves

The previous three chapters dealt with average decay for subsets of $\mathbb{R}^d$, which quickly reduces to the problem of average decay of Fourier transforms of measures supported on surfaces of co-dimension one. In this chapter we address the issue of Fourier transform of average decay of measures supported on curves in $\mathbb{R}^d$ with some tantalizing connection with the classical restriction theory.

8.1 Statement of results

Let Γ be a smooth (C^∞) immersed curve in $\mathbb{R}^d$ with parametrization $t \to \gamma(t)$ defined on a compact interval I and let $\chi \in C^\infty$ be supported in the interior of I. Let $\mu \equiv \mu_{\gamma,\chi}$ be defined by

$$\langle \mu, f \rangle = \int f(\gamma(t))\chi(t)dt \tag{8.1}$$

and define by

$$\widehat{\mu}(\xi) = \int \exp(-i\langle \xi, \gamma(t) \rangle)\chi(t)dt$$

its Fourier transform. For a large parameter R we are interested in the behavior of $\widehat{\mu}(R\omega)$ as a function on the unit sphere, in particular in the L^q norms

$$G_q(R) \equiv G_q(R;\gamma,\chi) := \Big(\int |\widehat{\mu}(R\omega)|^q d\omega \Big)^{\frac{1}{q}} \tag{8.2}$$

where $d\omega$ is the rotation invariant measure on S^{d-1} induced by Lebesgue measure in $\mathbb{R}^d$. The rate of decay depends on the number of linearly independent derivatives of the parametrization of Γ. Indeed if one assumes that for every t the derivatives $\gamma'(t), \gamma''(t), \dots, \gamma^{(d)}(t)$ are linearly independent, then from the standard van der Corput's lemma (see [180, p. 334]) one gets

$$G_\infty(R) = \max_\omega |\widehat{\mu}(R\omega)| = O(R^{-\frac{1}{d}}).$$

If one merely assumes that at most $d-1$ derivatives are linearly independent then one cannot in general expect a decay of $G_\infty(R)$; one simply considers curves which lie in a hyperplane. However Marshall [141] showed that one gets an optimal estimate for the L^2-*average decay*, namely

$$G_2(R) = O(R^{-\frac{1}{2}}) \tag{8.3}$$

as $R \to \infty$, for every compactly supported C^1 curve γ.

We are interested in estimates for the L^q-average decay, for $2 < q < \infty$. If γ is a straight line such extensions fail, and additional conditions are necessary. Our first result addresses the case of non-vanishing curvature.

Theorem 8.4. *Suppose that for all $t \in I$ the vectors $\gamma'(t)$ and $\gamma''(t)$ are linearly independent. Then for $R \geq 2$,*

(i)
$$G_q(R) \lesssim \begin{cases} R^{-\frac{1}{2}} (\log R)^{\frac{1}{2}-\frac{1}{q}} & \text{if } 2 \leq q \leq 4 \\ R^{-\frac{2}{q}} (\log R)^{\frac{1}{q}} & \text{if } 4 \leq q \leq \infty. \end{cases} \tag{8.5}$$

(ii) *Suppose that there is $N \in \mathbb{N}$ so that for every $\omega \in S^{d-1}$ the function $s \mapsto \langle \omega, \gamma''(s) \rangle$ changes sign at most N times on I. Then*

$$G_q(R) \lesssim R^{-\frac{1}{2}} \quad \text{if } 2 \leq q < 4 \tag{8.6}$$

and

$$G_q(R) \lesssim R^{-\frac{2}{q}} \quad \text{if } 4 < q \leq \infty. \tag{8.7}$$

Here and elsewhere the notation $a \lesssim b$ means $a \leq Cb$ for a suitable non-negative constant C.

The L^4 estimate

$$G_4(R) = O(R^{-\frac{1}{2}} [\log R]^{\frac{1}{4}})$$

is sharp even for nondegenerate curves, *cf.* Theorem 8.12 below. The estimate (8.6) is sharp and it is open whether for $q \neq 4$ there exists an example for which the logarithmic term in (8.5) is necessary.

The estimate (8.7) is sharp in the case where the curve lies in a two-dimensional subspace. Under stronger nondegeneracy assumptions this estimate can be improved. In particular one is interested in the case of nondegenerate curves in $\mathbb{R}^d$, meaning that for all t the vectors $\gamma^{(j)}(t)$, $j = 1, \dots, d$, are linearly independent. In the case $d = 2$ we have of course the optimal bound

$$G_q(R) = O(R^{-\frac{1}{2}})$$

for all $q \le \infty$, by the well-known stationary phase bound (for results for general curves in $\mathbb{R}^2$ and hypersurfaces in higher dimensions see [45] and references contained therein). The situation is more complicated for nondegenerate curves in higher dimensions, and Marshall [141] proved (essentially) optimal results for nondegenerate curves in $\mathbb{R}^d$ if $d = 3$ and $d = 4$.

We show that one gets close to optimal results for nondegenerate curves in all dimensions. Our method is different from the explicit computations in Marshall's paper and relies on a variable coefficient analogue of the Fourier restriction theorem due to Fefferman and Stein in two dimensions, see [71], and due to Drury [62] for curves in higher dimensions. The variable coefficient analogues are due to Carleson and Sjölin [51] (see also Hörmander [100]) in two dimensions and to Bak and Lee [9] in higher dimensions.

To formulate our result let, for $1 \le q \le \infty$,

$$\sigma_K(q) \equiv \sigma_K^d(q) = \begin{cases} \min_{\{k=2,\dots,d\}} \frac{1}{k} + \frac{k^2-k-2}{2kq}, & \text{for } K = d, \\ \min_{\{k=2,\dots,K\}} \left\{\frac{1}{k} + \frac{k^2-k-2}{2kq}, \frac{K}{q}\right\}, & \text{for } 2 \le K < d. \end{cases} \tag{8.8}$$

Theorem 8.9. *Suppose that for all $t \in I$ the vectors $\gamma'(t), \dots, \gamma^{(K)}(t)$ are linearly independent. Then for $R \ge 2$,*

$$G_q(R) \le C_\sigma R^{-\sigma}, \quad \sigma < \sigma_K(q). \tag{8.10}$$

For integers $k \ge 1$ set

$$q_k := \frac{k^2 + k + 2}{2} \tag{8.11}$$

so that $q_1 = 2$, $q_2 = 4$, $q_3 = 7$, $q_4 = 11$. Observe that the set of points $(q^{-1}, \sigma_d^d(q))$, $q \ge 2$, is the broken line joining the points

$$\left(\frac{1}{q_1}, \frac{1}{q_1}\right), \left(\frac{1}{q_2}, \frac{2}{q_2}\right), \dots, \left(\frac{1}{q_k}, \frac{k}{q_k}\right), \dots, \left(\frac{1}{q_{d-1}}, \frac{d-1}{q_{d-1}}\right), \left(0, \frac{1}{d}\right),$$

while for $K < d$, the set of points $(q^{-1}, \sigma_K^d(q))$ is the concave broken line joining the points

$$\left(\frac{1}{q_1}, \frac{1}{q_1}\right), \left(\frac{1}{q_2}, \frac{2}{q_2}\right), \dots, \left(\frac{1}{q_k}, \frac{k}{q_k}\right), \dots, \left(\frac{1}{q_K}, \frac{K}{q_K}\right), (0, 0).$$

Furthermore observe that $\sigma_K^d(q) > \frac{2}{q}$ if $3 \le K \le d$, and $q > 4$. The picture (see Figure 8.1) shows the graph $\{\frac{1}{q}, \sigma_K^K(q)\}$ as a function of $\frac{1}{q}$, for $K = 10$.

We emphasize that the graph of σ_K^d is slightly different for $d > K$, as then the left line segment connects (q_K^{-1}, Kq_K^{-1}) to $(0, 0)$.

Theorem 8.9 is sharp up only to endpoints, at least for nondegenerate curves (for which $\gamma'(t), \dots, \gamma^{(d)}(t)$ are linearly independent), and also for some other cases

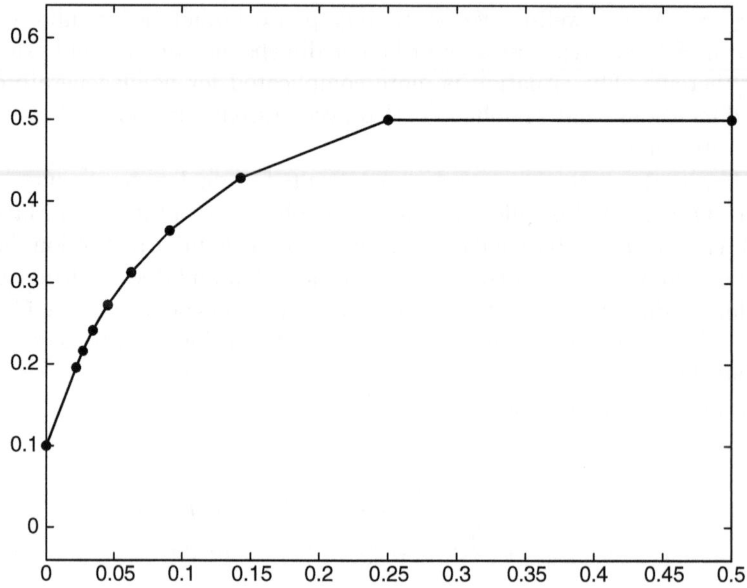

Figure 8.1: The graph $\left\{\frac{1}{q}, \sigma_{10}^{10}(q)\right\}$

where $K < d$, $\gamma'(t), \dots, \gamma^{(K)}(t)$ are independent and γ lies in a K-dimensional affine subspace. We note that for the case $K = d = 4$ Marshall [141] obtained the sharp bound

$$G_q(R) \lesssim R^{-\sigma_4^4(q)}$$

when $4 < q < 7$ and $q > 7$; moreover

$$G_q(R) \lesssim R^{-\sigma_4^4(q)} \log^{\frac{1}{q}}(R)$$

when $q = 4$ or $q = 7$ (the logarithmic term for the L^4 bound seems to have been overlooked in [141]).

We now state lower bounds for the average decay. The cutoff function χ is as in (8.1) (and $G_q(R)$ depends on χ).

Theorem 8.12. *Suppose that $2 \le K \le d$ and, for some $t_0 \in I$, the vectors $\gamma'(t_0), \dots, \gamma^{(K)}(t_0)$ are linearly independent. Then for suitable $\chi \in C_0^\infty$ there are $c > 0$, $R_0 \ge 1$ so that the following lower bounds hold for $R > R_0$.*

(i) *If $2 \le K \le d-1$ then*

$$G_q(R) \ge cR^{-\sigma_K(q)}, \quad 2 < q < q_K; \tag{8.13}$$

moreover

$$G_q(R) \ge cR^{-\sigma_K(q)} \log^{\frac{1}{q}}(R), \quad q \in \{q_k : k = 2, \dots, K-1\}. \tag{8.14}$$

(ii) *If $K = d$ then*

$$G_q(R) \geq cR^{-\sigma_d(q)}, \quad 2 < q \leq \infty, \tag{8.15}$$

$$G_q(R) \geq cR^{-\sigma_d(q)} \log^{\frac{1}{q}}(R), \quad q \in \{q_k : k = 2, \dots, d-1\}. \tag{8.16}$$

(iii) *If $2 \leq K \leq d-1$ and, in addition, $\gamma^{(K+1)} \equiv 0$ then*

$$G_q(R) \geq cR^{-\sigma_K(q)}, \quad 2 < q \leq \infty, \tag{8.17}$$

$$G_q(R) \geq cR^{-\sigma_K(q)} \log^{\frac{1}{q}}(R), \quad q \in \{q_k : k = 2, \dots, K\}. \tag{8.18}$$

Remark. A careful examination of the proof yields some uniformity in the lower bound. Assume that $\gamma^{(j)}(t_0) = e_j$, (the jth unit vector), $j = 1, \dots, K$, and $\|\gamma\|_{C^{K+3}(I)} \leq C_1$. Then there is $h = h(C_1) > 0$ so that for every smooth χ supported in $(-h, h)$ with $\Re\chi(t) > c_1 > 0$ in $(-\frac{h}{2}, \frac{h}{2})$ there exists an R_0 depending only on c_1, C_1, $\|\chi'\|_\infty$ and $\|\chi''\|_\infty$ so that the above lower bounds hold for $R \geq R_0$. We shall not pursue this point in detail.

Remark 8.19. Arkhipov, Chubarikov and Karatsuba [3], [4] proved sharp estimates for the $L^q(\mathbb{R}^d)$ norms of the Fourier transform of smooth densities on certain polynomial curves. The work of these authors shows that for, say

$$\gamma(t) = \sum_{k=1}^{d} t^k e_k, \quad t \in [0,1],$$

the Fourier transform $\widehat{d\sigma}$ belongs to $L^q(\mathbb{R}^d)$ if and only if

$$q > q_d = \frac{d^2 + d + 2}{2}.$$

This result seems to have been overlooked until recently; it rules out an $L^{q_d'}$ endpoint bound for the Fourier restriction problem associated to curves, *cf.* a discussion in [147] and a remark in [9]). More can be said in two dimensions where the endpoint restricted weak type $(\frac{4}{3})$ inequality for the Fourier restriction operator is known to fail by a Kakeya set argument, see [18]. We note that the lower bound in Chapter 2 of [4] is closely related to (8.15) and the method in [4] actually can be used to yield (8.15) for the curve $(t, \dots, t^d)$ in the range $q \geq q_{d-1}$; vice versa one notices $\sigma_d(q_d) = \frac{d}{q_d}$ and integrates the lower bound for $R^{d-1}G_q^q(R)$ in R to obtain lower bounds for $\|\widehat{d\sigma}\|_{L^q(\mathbb{R}^d)}$.

A variant of an argument in [4] can be shown to close the ε gap between upper and lower bounds in some cases. We formulate one such result.

Theorem 8.20. *Suppose that γ is smooth and is either of finite type, or polynomial.*

Assume that $\gamma'(t), \dots, \gamma^{(K)}(t)$ are linearly independent, for every $t \in I$. Then the following holds:

(i) *If* $K = d$ *then*

$$G_q(R) \le C_q R^{-\sigma_d(q)}, \quad q \ge 2, \quad q \notin \{q_k : k = 2, \dots, d-1\}, \tag{8.21}$$

and

$$G_q(R) \lesssim R^{-\sigma_d(q)} \log^{\frac{1}{q}}(R), \quad q \in \{q_k : k = 2, \dots, d-1\}. \tag{8.22}$$

(ii) *If* $2 \le K \le d-1$ *then*

$$G_q(R) \le C_q R^{-\sigma_K(q)}, \quad q \ge 2, \quad q \notin \{q_k : k = 2, \dots, K\}. \tag{8.23}$$

and

$$G_q(R) \lesssim R^{-\sigma_K(q)} \log^{\frac{1}{q}}(R), \quad q \in \{q_k : k = 2, \dots, K\} \tag{8.24}$$

It is understood that the implicit constants in (8.21) and (8.23) depend on q as $q \to q_k$. Note that in the finite type case (8.23) and (8.24) can be improved for $q > q_K$ since we have some nontrivial decay for $G_\infty(R)$. However, for the sharpness in the most degenerate case compare Theorem 8.12, part (iii).

The result of Theorem 8.20, for polynomial curves, could be used to obtain the upper bounds of Theorem 8.9, which involves a loss of R^ε, by a polynomial approximation argument. Note however, that such an argument requires upper bounds for derivatives of γ up to order $C + \varepsilon^{-1}$, as $\varepsilon \to 0$. An examination of the proof of Theorem 8.9 shows that one can get away with upper bounds for the derivatives up to order N where N depends on the dimension but not on ε.

8.2 Upper bounds, I

We shall now prove part (ii) of Theorem 8.4, (*i.e.*, (8.6), (8.7)) under the less restrictive smoothness condition $\gamma \in C^2(I)$; we recall the assumptions that $\gamma'(t)$ and $\gamma''(t)$ are linearly independent and that we also require that the functions $s \mapsto \langle \omega, \gamma''(s) \rangle$ have at most a bounded number of sign changes on I. Note that this hypothesis is certainly satisfied if γ is a polynomial, or a trigonometric polynomial, or smooth and of finite type.

We need a result on oscillatory integrals which is a consequence of the standard van der Corput Lemma; it is also related to a more sophisticated statement on oscillatory integrals with polynomial phases in [153].

Let η be a C^∞ function with support in $(-1, 1)$ so that $\eta(s) = 1$ in $(-\frac{1}{2}, \frac{1}{2})$; we also assume that η' has only finitely many sign changes. Let $\eta_1(s) = \eta(s) - \eta(2s)$ (so that $\frac{1}{4} \le |s| \le 1$ on $\operatorname{supp} \eta_1$) and let

$$\eta_l(s) = \eta_1(2^{l-1}s)$$

so that $2^{-l-1} \le |s| \le 2^{-l+1}$ on the support of η_l.

Lemma 8.25. *Let I be a compact interval and let $\chi \in C^1(I)$. Let $\phi \in C^2(I)$ and suppose that ϕ'' changes signs at most N times in I.*

Then, for $1 \leq 2^l \leq \lambda$,

$$\left|\int_I \eta_l(\phi'(s))e^{i\lambda\phi(s)}\chi(s)ds\right| \leq CN2^l\lambda^{-1}.$$

Proof. We may decompose I into subintervals J_i, $1 \leq i \leq K$, $K \leq 2N+2$, so that both ϕ' and ϕ'' do not change sign in each J_i. Each interval J_i can be further decomposed into a bounded number of intervals $J_{i,k}$ so that $\eta'(\phi')$ is of constant sign in $J_{i,k}$. It suffices to estimate the integral $\mathcal{I}_{i,k}$ over $J_{i,k}$. By the standard van der Corput Lemma, the bound $\mathcal{I}_{i,k} = O(\frac{2^l}{\lambda})$ follows if we can show that

$$\int_{J_{i,k}} \left|\frac{\partial}{\partial s}\left(\eta_l(\phi'(s))\chi(s)\right)\right| ds \leq C$$

which immediately follows from

$$\int_{J_{i,k}} \left|2^l\phi''(s)\eta_1'(2^l\phi'(s))\right|ds \leq C. \tag{8.26}$$

But by our assumption on the signs of ϕ', ϕ'', and η' the left-hand side is equal to

$$\left|\int_{J_{i,k}} 2^l\phi''(s)\eta_1'(2^l\phi'(s))ds\right| = \left|\int_{J_{i,k}} \frac{\partial}{\partial s}\Big(\eta_l(\phi'(s))\Big)ds\right| \leq C. \qquad \square$$

Proof of (8.6) *and* (8.7). We may assume that Γ is parametrized by arclength and that the support of χ is small (of diameter $\lesssim 1$). Determine the integer $M(R)$ by $2^M \leq R < 2^{M+1}$. With η_l as above define for $l < M$,

$$g_{R,l}(\omega) = \int e^{iR\langle\omega,\gamma(s)\rangle}\chi(s)\eta_l(\langle\omega,\gamma'(s)\rangle)ds$$

and for $l = M$ define $g_{R,M}$ similarly by replacing the cutoff $\eta_l(\langle\omega,\gamma'(s)\rangle)$ with $\eta(2^M\langle\omega,\gamma'(s)\rangle)$. We can decompose

$$\int e^{iR\langle\omega,\gamma(t)\rangle}\chi(t)dt = \sum_{l\leq M} g_{R,l}(\omega)$$

and observe that $g_{R,l} = 0$ if $l \leq -C$.

It follows from Lemma 8.25 that

$$\sup_{\omega\in S^{d-1}} |g_{R,l}(\omega)| \lesssim \frac{2^l}{R}. \tag{8.27}$$

We also claim that

$$\Big(\int |g_{R,l}(\omega)|^2 d\omega\Big)^{\frac{1}{2}} \lesssim \begin{cases} 2^l R^{-1} & \text{if } 2^l \leq R^{\frac{1}{2}}, \\ 2^{-l}(1+\log(2^{2l}R^{-1}))^{\frac{1}{2}} & \text{if } 2^l \geq R^{\frac{1}{2}}. \end{cases} \tag{8.28}$$

Given (8.27) and (8.28) we deduce that

$$\|g_{R,l}\|_{L^q(S^{d-1})} \le \|g_{R,l}\|_{L^2(S^{d-1})}^{\frac{2}{q}} \|g_{R,l}\|_{L^\infty(S^{d-1})}^{1-\frac{2}{q}} \lesssim \begin{cases} 2^l R^{-1} & \text{if } 2^l \le R^{\frac{1}{2}}, \\ 2^{l(1-\frac{4}{q})}(1+\log(2^{2l}R^{-1}))^{\frac{1}{q}} R^{-1+\frac{2}{q}} & \text{if } 2^l \ge R^{\frac{1}{2}}. \end{cases} \tag{8.29}$$

If $q \neq 4$ the asserted bound $O(R^{-\frac{2}{q}})$ bound follows by summing in l.

We now turn to the proof of (8.28). Note that (8.28) follows immediately from (8.27) if $2^l \le R^{\frac{1}{2}}$. Now let $2^l \ge R^{\frac{1}{2}}$. For the L^2 estimate in this range we shall just use the non-vanishing curvature assumption on Γ. We need to estimate the L^2 norm of $g_{R,l}$ over a small coordinate patch $\mathcal{V}$ on the sphere where we use a regular parametrization $y \to \omega(y)$, $y \in [-1,1]^{d-1}$; i.e.,

$$\left| \int u(y) g_{R,l}(\omega(y)) \overline{g_{R,l}(\omega(y))} dy \right| \lesssim 2^{-2l}(1+\log(2^{2l}R^{-1})), \tag{8.30}$$

where $u \in C_0^\infty$, so that $\omega(y) \in \mathcal{V}$ if $y \in \text{supp}(u)$. The left-hand side of (8.30) can be written as

$$\mathcal{I}_l := \iint_{s_1,s_2} \int_y u(y) e^{iR\langle \omega(y), \gamma(s_1)-\gamma(s_2)\rangle} \chi(s_1)\overline{\chi(s_2)} \times \eta(2^l\langle \omega(y), \gamma'(s_1)\rangle)\eta(2^l\langle \omega(y), \gamma'(s_2)\rangle) dy ds_1 ds_2$$

and we note that on the support of the amplitude we get that $\gamma'(s_i)$ is almost perpendicular to ω, *i.e.*, we may assume by the assumption of small supports that there is a direction $w = (w_1, \dots, w_{d-1})$ so that

$$\left| \sum_{\nu=1}^{d-1} w_\nu \frac{\partial}{\partial y_\nu} \langle \omega(y), \gamma'(s)\rangle \right| \ge \frac{1}{2}$$

if $s \in \text{supp}(\chi)$ and $y \in \text{supp}(u)$. By a rotation in parameter space we may assume that

$$\left| \frac{\partial}{\partial y_1} \langle \omega(y), \gamma'(s)\rangle \right| \ge \frac{1}{2} \quad \text{if } s \in \text{supp}(\chi), \quad y \in \text{supp}(u). \tag{8.31}$$

Now let, for fixed unit vectors v_1, v_2 and $\delta > 0$,

$$\mathcal{U}_\delta(v_1, v_2) = \{\omega \in S^{d-1} : |\langle \omega, v_1\rangle| \le \delta, |\langle \omega, v_2\rangle| \le \delta\}$$

and observe that the spherical measure of this region is at most $O(\delta)$; moreover this bound can be improved if $|v_1 - v_2|$ is $\ge \delta$. Namely if $\alpha(v_1, v_2)$ is the acute angle between v_1 and v_2 then

$$\text{meas}(\mathcal{U}_\delta(v_1, v_2)) \lesssim \min\left\{\delta, \frac{\delta^2}{\sin\alpha(v_1, v_2)}\right\}. \tag{8.32}$$

The condition (8.31) implies that

$$\left|\left\langle \frac{\partial}{\partial y_1}\omega(y), \gamma(s_1) - \gamma(s_2)\right\rangle\right| \geq c|s_1 - s_2|$$

and given the regularity of the amplitude we can gain by a multiple integration by parts in y_1 provided that $|s_1 - s_2| \geq \frac{2^l}{R}$; indeed we gain a factor of

$$O(R^{-1}|s_1 - s_2|^{-1}2^l)$$

with each integration by parts. We obtain, for any N,

$$\begin{aligned} \mathcal{I}_l \lesssim \int_{s_1 \in \text{supp}(\chi)} \Bigg[\int_{|s_2 - s_1| \leq c} & \text{meas}\,(\mathcal{U}_{2^{-l}}(\gamma'(s_1), \gamma'(s_2))) \\ & \times \min\{1, (R|s_1 - s_2|2^{-l})^{-N}\}\, ds_2\Bigg] ds_1. \end{aligned} \tag{8.33}$$

By the assumption that $|\gamma''(s)|$ is bounded below and γ' and γ'' are orthogonal we get as a consequence of (8.32)

$$\text{meas}\,(\mathcal{U}_{2^{-l}}(\gamma'(s_1), \gamma'(s_2))) \leq \min\{2^{-l}, 2^{-2l}|s_1 - s_2|^{-1}\}.$$

Now we use this bound and integrate out the s_2 integral in (8.33) and see that the main contribution comes from the region where

$$2^{-l} \leq |s_1 - s_2| \leq \frac{2^l}{R}$$

which yields the factor $\log(R2^{-2l})$ in (8.30). □

An application. We consider a C^2 curve $\gamma : [-a, a] \to \mathbb{R}^d$ with non-vanishing curvature and assume that, as in (8.6), the function $s \mapsto \langle \omega, \gamma''(s)\rangle$ has a bounded number of sign changes.

Let μ be the measure induced by the Lebesgue measure on Γ, multiplied by a smooth cutoff function. For every $\sigma \in SO\,(d)$ define μ_σ by $\mu_\sigma\,(E) = \mu\,(\sigma E)$ and for every test function f in $\mathbb{R}^d$,

$$Tf\,(x, \sigma) = f * \mu_\sigma\,(x)\,.$$

We are interested in the

$$L^p(\mathbb{R}^d) \to L^s(SO(d), L^q(\mathbb{R}^d))$$

mapping properties, in particular for $q = p' = \frac{p}{p-1}$. This question had been investigated in [163] for curves in the plane, with essentially sharp results in this case, see also [43]. The standard example, namely testing T on characteristic functions of balls of small radius, yields the necessary condition $1+\frac{d-1}{q}\geq\frac{d}{p}$. Setting $q=p'$ we see that the

$$L^p(\mathbb{R}^d)\to L^s(SO(d),L^{p'}(\mathbb{R}^d))$$

fails for $p<\frac{2d-1}{d}$ (independent of s).

The approach in [163] together with the inequality (8.6) yields

$$\|Tf\|_{L^s(SO(d),L^{p'}(\mathbb{R}^d))}\leq C_p\|f\|_p,\quad p=\tfrac{2d-1}{d},\quad s<\tfrac{4d-2}{d}. \tag{8.34}$$

Proof of (8.34). We imbed T in an analytic family of operators. After rotation and reparametrization (modifying the cut-off function) we may assume that

$$\gamma(t)=\sum_{j=1}^{d-1}\varphi_j(t)e_j+te_d,$$

with $\varphi_j(0)=0$. Let $z\in\mathbb{C}$ such that $\operatorname{Re}z>0$ and define a distribution i_z by

$$\langle i_z,\chi\rangle=(\Gamma(z))^{-1}\int_0^{+\infty}\chi(t)\,t^{z-1}dt.$$

Then define μ_σ^z by

$$\widehat{\mu_\sigma^z}(\xi)=\widehat{\mu_\sigma}(\xi)\prod_{j=1}^{d}\widehat{i_z}(\langle\sigma\xi,e_j\rangle)$$

and T^z by $T^zf(x,\sigma)=\mu_\sigma^z*f$. Following [163] one observes that $\mu_\sigma^{1+i\lambda}$ is a bounded function, namely we have

$$|\langle\mu^{1+i\lambda},g\rangle|\lesssim\int_{\mathbb{R}}\int_{\mathbb{R}^{d-1}}\left|g\left(x_de_d+\sum_{j=1}^{d-1}(y_j+\phi_j(x_d)e_j\right)\right|dy_1\cdots dy_{d-1}dx_d\lesssim\|g\|_1$$

so that

$$T^{1+i\lambda}:L^1(\mathbb{R}^d)\to L^\infty(SO(d)\times\mathbb{R}^d). \tag{8.35}$$

We also have

$$T^{-\frac{1}{2d-2}+i\lambda}:L^2(\mathbb{R}^d)\to L^q(SO(d),L^2(\mathbb{R}^d)),\quad 2\leq q<4. \tag{8.36}$$

The implicit constants in both inequalities are at most exponential in λ. Thus we obtain the assertion (8.34) by analytic interpolation of operators.

To see (8.36), we observe that

$$\widehat{i_z}(\tau)=O\big(|\tau|^{-\operatorname{Re}(z)}\big)$$

and apply Plancherel's theorem and then Minkowski's integral inequality to bound for $\alpha > 0$,

$$\begin{aligned}
&\|T^{-\alpha+i\lambda} f\|_{L^q(L^2)}^2 \\
&= \left(\int_{SO(d)} \left(\int_{\mathbb{R}^d} |\widehat{f}(\xi)\widehat{\mu}_\sigma|^2 \prod_{j=1}^{d-1} |\widehat{i_{-\alpha+i\lambda}}(\langle \sigma\xi, e_j \rangle)|^2 d\xi \right)^{\frac{q}{2}} d\sigma \right)^{\frac{2}{q}} \\
&\lesssim \int_{\mathbb{R}^d} \left|\widehat{f}(\xi)\right|^2 \left(\int_{SO(d)} \left|\widehat{\mu_\sigma}(\xi)\right|^q \prod_{j=1}^{d-1} |\langle \sigma\xi, e_j \rangle|^{\alpha q} d\sigma \right)^{\frac{2}{q}} d\xi,
\end{aligned}$$

and by (8.6) and the assumption $q < 4$ the last expression is dominated by a constant times

$$\begin{aligned}
&\int_{\mathbb{R}^d} \left|\widehat{f}(\xi)\right|^2 |\xi|^{2\alpha(d-1)} \left(\int_{SO(d)} \left|\widehat{\mu_\sigma}(\xi)\right|^q d\sigma \right)^{\frac{2}{q}} d\xi \\
&\qquad \lesssim \int_{\mathbb{R}^d} \left|\widehat{f}(\xi)\right|^2 |\xi|^{2\alpha(d-1)-1} d\xi.
\end{aligned}$$

For $\alpha = (2d-2)^{-1}$, this yields the bound (8.36). □

Remark. We do not know whether the index $s = \frac{4d-2}{d-1}$ in (8.34) is sharp. The following example only shows that we need $s \leq 10$ for $d = 3$. Let $\gamma(t) = (t, t^2, 0)$ and let χ_{B_δ} be a box centered at the origin with sides parallel to the axes and having side lengths $1, 1$ and δ. A computation shows that $|T\chi_{B_\delta}(x, \sigma)| \geq c$ for σ in a set of measure ε^2 and x in a set of measure ε, for some small $\varepsilon > 0$. It follows that $p^{-1} \leq 2s^{-1} + q^{-1}$. For $p = \frac{5}{3}$ and thus $p' = \frac{5}{2}$ this yields $s \leq 10$.

8.3 Upper bounds, II

We are now concerned with the proof of Theorem 8.9 and the proof of part (i) of Theorem 8.4. For the latter we use a version of the Carleson–Sjölin theorem ([51], [100]), and for Theorem 8.9 we use a recent generalization due to Bak and Lee [9]. These we now recall.

Consider, for large positive R,

$$T_R f(x) = \int_{\mathbb{R}} e^{iR\phi(x,t)} a(x, t) f(t)\, dt$$

with real-valued phase function $\phi \in C^\infty(\mathbb{R}^n \times \mathbb{R})$, and compactly supported amplitude $a \in C_0^\infty(\mathbb{R}^n \times \mathbb{R})$. Assume the non-vanishing torsion condition

$$\det \left(\frac{\partial}{\partial t} (\nabla_x \phi), \frac{\partial^2}{\partial t^2} (\nabla_x \phi), \ldots, \frac{\partial^n}{\partial t^n} (\nabla_x \phi) \right) \neq 0 \tag{8.37}$$

on the support of a. Then if

$$\frac{1}{p}+\frac{n(n+1)}{2q}=1$$

and $q>\frac{n^2+n+2}{2}$, there is a constant C_q independent of f and of $R\geq 2$ such that

$$\|T_R f\|_{L^q(\mathbb{R}^n)}\leq C_q R^{-\frac{n}{q}}\|f\|_{L^p(\mathbb{R})}. \tag{8.38}$$

When $n=2$ it is well known that a slight modification of Hörmander's proof ([100]) of the Carleson–Sjölin theorem gives the endpoint result

$$||T_R f||_{L^4(\mathbb{R}^n)}\lesssim R^{-\frac{1}{2}}\log^{\frac{1}{4}}R\,||f||_{L^4(\mathbb{R})}\,; \tag{8.39}$$

see also [150], where a somewhat harder vector-valued analogue is proved.

In order to establish estimates (8.5) we need to show that, under the assumption of linear independence of $\gamma'(t)$ and $\gamma''(t)$ (for each $t\in I$),

$$G_4(R)\lesssim R^{-\frac{1}{2}}[\log R]^{\frac{1}{4}}. \tag{8.40}$$

To establish (8.10) under the assumption that the first K derivatives are linearly independent for every $t\in I$, we need to show that for any $2\leq k\leq K$,

$$G_q(R)\lesssim R^{-\frac{k}{q}},\quad q>q_k, \tag{8.41}$$

where q_k is as in (8.11). All other estimates in (8.5), (8.10) follow by the usual convexity property of the L^p norm, *i.e.*,

$$\|F\|_p\leq\|F\|_{p_0}^{1-\vartheta}\|F\|_{p_1}^{\vartheta}\qquad\text{for}\qquad\frac{1}{p}=\frac{1-\vartheta}{p_0}+\frac{\vartheta}{p_1}.$$

Proof of (8.41) *and* (8.40). Let

$$F_R(\omega)=\int_{\mathbb{R}}e^{i\langle\omega,\gamma(t)\rangle}a(\omega)\chi(t)dt. \tag{8.42}$$

By compactness, we can suppose that χ is supported in $(-\varepsilon,\varepsilon)$, with ε as small as we need. Divide the sphere S^{d-1} into two subsets A and B; here in A, the unit normal to the sphere is essentially orthogonal to the span of the vectors $\gamma'(0),\dots,\gamma^{(k)}(0)$, and in B, the unit normal to the sphere is close to the span of $\gamma'(0),\dots,\gamma^{(k)}(0)$.

Now consider a coordinate patch V of diameter ε on A and parametrize it by $y\mapsto\omega(y)$ with $y\in\mathbb{R}^{d-1}$, y near Y_0. From the defining property of A, it follows that the vectors $\nabla_y\langle\omega(\cdot),\gamma^{(j)}(t)\rangle$, $j=1,\dots,k$ are linearly independent when evaluated at y near Y_0, provided that $|t|<\varepsilon$. Therefore we can choose the parameterization

$$y=(x',y'')=(x_1,\dots,x_k,y_{k+1},\dots,y_{d-1})$$

in such a way that also the vectors

$$\nabla_{x'}\langle\omega(\cdot),\gamma^{(j)}(t)\rangle,\quad j=1,\dots,k,$$

are linearly independent.

If we consider y'' as a parameter and we define

$$\phi^{y''}(x',t)=\langle\omega(x',y''),\gamma(t)\rangle,$$

then the phase functions $\phi^{y''}$ satisfy condition (8.37) uniformly in y''. We also have upper bounds for the higher derivatives in ω and γ which are uniform in y'' as well (here y'' is taken from a relevant compact set). Thus one can apply the Bak-Lee result (8.38) in k dimensions to obtain, for fixed y'',

$$\Big(\int\Big|\int_{\mathbb{R}} e^{iR\phi^{y''}(y',t)}a(\omega(y',y''))\chi(t)dt\Big|^q dy'\Big)^{\frac{1}{q}} \lesssim R^{-\frac{k}{q}},\quad q>q_k. \tag{8.43}$$

An integration in y'' yields

$$\|F_R\|_{L^q(V)}\lesssim R^{-\frac{k}{q}}$$

for $q>\frac{k^2+k+2}{2}$. Similarly if $k=2$ and $q=4$ we can apply (8.39) in two variables to obtain

$$\|F_R\|_{L^4(V)}\lesssim R^{-\frac{1}{2}}\log^{\frac{1}{4}}R.$$

This settles the main estimate for the $L^q(A)$ norm. As for contribution of the $L^q(B)$ norm we recall that the unit normal to the sphere is close to the span of $\gamma'(0),\dots,\gamma^{(k)}(0)$, and thus

$$\sum_{j=1}^{k}|\langle\omega,\gamma^{(j)}(t)\rangle|>0.$$

Therefore we can apply van der Corput's lemma and obtain the L^∞ estimate

$$\|F_R\|_{L^\infty(B)}\lesssim R^{-\frac{1}{k}}. \tag{8.44}$$

For $k=2$, this completes the proof of the theorem. For $2<k\le K$, we argue by induction. We assume that the asserted estimate holds for $k-1$, $(k\ge 3)$; that is

$$\|F_R\|_q\lesssim R^{-\frac{k-1}{q}}\qquad\text{for}\qquad q>q_{k-1}=\tfrac{k^2-k+2}{2} \tag{8.45}$$

where the implicit constants depend on q. If $\vartheta_k=1-\frac{q_{k-1}}{q_k}$ then we use the relation $q_k-q_{k-1}=k$ to verify that

$$\frac{(1-\vartheta_k)(k-1)}{q_{k-1}}+\frac{\vartheta_k}{k}=\frac{k}{q_k}.$$

Thus by a convexity argument we see that a combination of (8.44) and (8.45) yields that

$$\|F_R\|_{L^q(B)}\lesssim R^{-\frac{k}{q}},\qquad\text{for}\qquad q>\tfrac{k^2+k+2}{2}=q_k.$$

Together with the corresponding bound for $\|F_R\|_{L^q(A)}$ proved above, this concludes the proof. □

8.4 Upper bounds, III

We give the proof of Theorem 8.20 under the *finite type assumption.* By compactness, there is an integer $L \geq d$ and a constant $c > 0$ so that for every $s \in I$ and every $\theta \in S^{d-1}$ we have

$$\sum_{n=1}^{L} |\langle \gamma^{(n)}(s), \theta \rangle| \geq c.$$

We shall argue by induction on k. By Theorem 8.9 the conclusion holds for $k = 2$. Assume $k > 2$, and that the desired inequalities are already proved for $2 \leq q \leq q_{k-1}$.

Let F_R be as in (8.42) and assume that the cut-off function χ is supported in $(-\varepsilon, \varepsilon)$. As in the proof of Theorem 8.9 we split the sphere into subsets A and B where in A, the unit normal to the sphere is almost perpendicular to the span of the vectors $\gamma'(t), \dots, \gamma^{(k)}(t)$, for all $|t| < \varepsilon$ and in B, the projections of the unit normals to the sphere to the span of $\gamma'(t), \dots, \gamma^{(k)}(t)$ have length $\geq c > 0$.

We shall estimate the $L^q(A \cap \Omega)$ norm of F_R on a small patch Ω on the sphere, and by further localization we may assume by the finite type assumption that there is an $n \leq L$ so that

$$|\langle \gamma^{(n)}(s), \theta \rangle| \geq c > 0, |s| \leq \varepsilon, \theta \in \Omega. \tag{8.46}$$

We distinguish between the case $n \geq k$ and $n < k$. First we assume $n \geq k$ (the main case). Then there is the pointwise bound

$$F_R(\theta) \lesssim \min \{1, H_R(\theta))^{-1}\} \tag{8.47}$$

where

$$H_R(\theta) = \min_{s \in I} \max_{1 \leq j \leq n} R^{\frac{1}{j}} |\langle \gamma^{(j)}(s), \theta \rangle|^{\frac{1}{j}}.$$

This is immediate from van der Corput's Lemma; indeed the finite type assumption allows the decomposition of the interval $[-\varepsilon, \varepsilon]$ into a bounded number of subintervals so that on each subinterval all derivatives of

$$s \mapsto \langle \gamma^{(j)}(s), \theta \rangle, \quad 1 \leq j \leq n-1,$$

are monotone and of the same sign. We now have to estimate the $L^q(A \cap \Omega)$ norm of the right-hand side of (8.47).

For an $l > 0$ consider the set

$$\Omega_l(R) = \{\theta \in \Omega : H_R(\theta) \in [2^l, 2^{l+1})\}. \tag{8.48}$$

By (8.46) we have $|H_R(\theta)| \gtrsim R^{\frac{1}{n}}$. Thus only the values with

$$2^l \gtrsim R^{\frac{1}{n}} \tag{8.49}$$

are relevant (and likewise the set of $\theta \in \Omega$ for which $H_R(\theta) \lesssim 1$ is empty if R is large).

By the definition of H_R we can find a point $s_* = s_*(\theta)$ and an integer j_*, $1 \le j^* \le n$, so that

$$H_R(\theta) = |R\langle \gamma^{(j_*)}(s_*), \theta\rangle|^{\frac{1}{j_*}}$$

and

$$|R\langle \gamma^{(j)}(s_*), \theta\rangle|^{\frac{1}{j}} \le H_R(\theta)$$

for all $\theta \in \Omega$ and all $j \le n$. This implies

$$|\langle \gamma^{(j)}(s_*), \theta\rangle| \lesssim 2^{(l+1)j} R^{-1}, \text{ if } \theta \in \Omega_l(R), \quad j \le n. \tag{8.50}$$

We shall now apply a nice idea of [3]: We divide our interval $(-\varepsilon, \varepsilon)$ into $O(2^l)$ intervals $I_{\nu,l}$ of length $\approx 2^{-l}$, with right endpoints t_ν, so that $t_\nu - t_{\nu-1} \approx 2^{-l}$. The point s_* lies in one of these intervals, say in I_{ν_*}. We estimate $|\langle \gamma^{(j)}(t_{\nu_*}), \theta\rangle|^{\frac{1}{j}}$ in terms of $H_R(\theta)$. By a Taylor expansion we get

$$\langle \gamma^{(j)}(t_{\nu_*}), \theta\rangle = \sum_{r=0}^{n-j-1} \langle \gamma^{(j+r)}(t_{\nu_*}), \theta\rangle \frac{(t_{\nu_*} - s_*)^r}{r!} + \langle \gamma^{(n)}(\widetilde{t}), \theta\rangle \frac{(t_{\nu_*} - s_*)^{n-j}}{(n-j)!} \tag{8.51}$$

where $\widetilde{t}$ is between s_* and t_{ν_*}. By (8.50) the terms in the sum are all $O(2^{lj}R^{-1})$. The remainder term is $O(2^{-l(n-j)})$ which is also $O(2^{lj}R^{-1})$, by the condition (8.49).

Now define

$$\Omega_{\nu,l} = \{\theta \in \Omega : |\langle \gamma^{(j)}(t_\nu), \theta\rangle| \le C 2^{lj} R^{-1}, j = 1, \ldots, n\}$$

and if C is sufficiently large then the set $\Omega_l(R)$ is contained in the union of the sets $\Omega_{\nu,l}$; the constant C can be chosen independently of l and R.

In view of the linear independence of the vectors $\gamma^{(j)}(t_\nu)$, $j = 1, \ldots, k$ and the condition $\theta \in A$, the measure of the set $\Omega_{\nu,l}$ is

$$O\left(\prod_{s=1}^{k} (2^{sl} R^{-1})\right) = O(2^{\frac{lk(k+1)}{2}} R^{-k}),$$

for every $1 \le \nu \lesssim 2^l$, and thus the measure of the set $\Omega_l(R)$ is

$$O(2^{\frac{l(k^2+k+2)}{2}} R^{-k}).$$

On $\Omega_l(R)$ we have

$$|F_R(\theta)| \le H_R(\theta)^{-1} \lesssim 2^{-l}.$$

Therefore

$$\int_{\Omega \cap A} |F_R(\theta)|^q d\theta \lesssim \sum_{cR^{\frac{1}{n}} \le 2^l \le cR} 2^{-lq} 2^{\frac{l(k^2+k+2)}{2}} R^{-k}, \tag{8.52}$$

which yields the endpoint bound

$$\Big(\int_{\Omega\cap A} |F_R(\theta)|^{q_k} d\theta\Big)^{\frac{1}{q_k}} \lesssim R^{-\frac{k}{q_k}} (\log R)^{\frac{1}{q_k}}.$$

Of course we also get (by using the same argument with just $k-1$ derivatives)

$$\int_\Omega |F_R(\theta)|^q d\theta \lesssim \sum_{cR^{\frac{1}{n}} \le 2^l \le cR} 2^{-lq} 2^{\frac{l(k^2-k+2)}{2}} R^{1-k} \tag{8.53}$$

which yields the sharp $L^{q_{k-1}}(A\cap\Omega)$ bound. Now we consider q satisfying $q_{k-1} < q < q_k$, $k < d$ or $q_{d-1} < q < \infty$ and $K = d$. We distinguish the cases (i) $2^l < R^{\frac{1}{k}}$ and (ii) $2^l \ge R^{\frac{1}{k}}$. In the first case we use (8.52) while in the second case we use (8.53). Then in the case $k < d$,

$$\Big(\int_\Omega |F_R(\theta)|^q d\theta\Big)^{\frac{1}{q}} \lesssim R^{-k}\Big(\sum_{2^l < R^{\frac{1}{k}}} 2^{\frac{l(-2q+k^2+k+2)}{2}} + \sum_{2^l \ge R^{\frac{1}{k}}} 2^{\frac{l(-2q+k^2-k+2)}{2}} R\Big)^{\frac{1}{q}}$$

which is bounded by

$$CR^{-\frac{1}{k}-\frac{k^2-k-2}{2kq}}$$

if $q_{k-1} < q < q_k$. If $K = k = d$ then only values with $2^l \ge R^{\frac{1}{d}}$ are relevant and only the second sum in the last displayed line occurs. Thus if $K = d$ we obtain the estimate

$$CR^{-\frac{1}{d}-\frac{d^2-d-2}{2dq}}$$

for $q > q_{d-1}$.

Now if $n < k$ one gets even better bounds; we use the induction hypothesis. First note that for $n = 1, 2$ integration by parts, or van der Corput's lemma, yields a better bound; therefore assume $n \ge 3$. We have the bounds

$$\|F_R\|_{L^\infty(\Omega)} \lesssim R^{-\frac{1}{n}} \qquad \text{and} \qquad \|F_R\|_{L^{q_n}(\Omega)} \lesssim R^{-\frac{n}{q_n}} (\log R)^{\frac{1}{q_n}};$$

the first one by van der Corput's Lemma and the second one by the induction hypothesis. By convexity this yields

$$\|F_R\|_{L^{q_k}} \lesssim R^{-\alpha(k,n)} \log R^{\frac{1}{q_k}}, \qquad \text{where} \qquad \alpha(k,n) = \frac{n}{q_k} + \frac{1-\frac{q_n}{q_k}}{n},$$

and one checks that

$$\alpha(k,n) = \frac{k}{q_k} + \frac{(k-n)(k+1-n)}{2nq_k}$$

if $n < k$, so that one gets a better estimate. The case $q_{k-1} < q < q_k$, $n < k$, is handled in the same way. This yields the desired bounds for the $L^q(A)$ norm of F_R.

For the $L^q(B)$ bound we may use van der Corput's estimate with $\le k$ derivatives to get an L^∞ bound $O(R^{-\frac{1}{k}})$; we interpolate this with the appropriate L^p bound for $q_{k-2} < p \le q_{k-1}$ which holds by the induction hypothesis; the argument is similar to that in the proof of Theorem 8.9. This finishes the argument under the finite type assumption.

Modification for polynomial curves. If the coordinate functions γ_j are polynomials of degree $\le L$ we need to take $n = L$ in the definition of $H_R(\theta)$. We use, for the case $l > 0$, the analogue of the Taylor expansion (8.51) up to order L with zero remainder term (again $n = L$). As above we obtain for $l > 0$ the bound

$$\int_{\Omega_l(R)} |F_R(\theta)|^q d\theta \lesssim 2^{-lq} R^{-k} \min\left\{2^{\frac{l(k^2+k+2)}{2}}, 2^{\frac{l(k^2-k+2)}{2}} R\right\}.$$

Summing in $l > 0$ works as before. However we also have a contribution from the set

$$\Omega_0(R) = \{\theta \in \Omega : H_R(\theta) \le 1\}.$$

By the polynomial assumption a Taylor expansion (now about the point s_*, without remainder) is used to show that $\Omega_0(R)$ is contained in the subset of A where

$$|\langle \gamma^{(j)}(s_*), \theta\rangle| \le CR^{-1}, \quad j = 1, \dots, k.$$

This set has measure $O(R^{-k})$. Thus the desired bound for $l = 0$ follows as well. □

8.5 Asymptotics for oscillatory integrals revisited

We examine the behavior of some known asymptotics for oscillatory integrals under small perturbations. This will be used in the subsequent section to prove the lower bounds of Theorem 8.12.

For $k = 2, 3, \dots,$ there is the following formula for $\lambda > 0$:

$$\int_{-\infty}^{\infty} e^{i\lambda s^k} ds = \alpha_k \lambda^{-\frac{1}{k}}, \tag{8.54}$$

where

$$\alpha_k = \begin{cases} \frac{2}{k}\Gamma\left(\frac{1}{k}\right) \sin\left(\frac{(k-1)\pi}{2k}\right), & k \text{ odd}, \\ \frac{2}{k}\Gamma\left(\frac{1}{k}\right) \exp\left(i\frac{\pi}{2k}\right), & k \text{ even}. \end{cases} \tag{8.55}$$

(8.54) is proved by standard contour integration arguments and implies asymptotic expansions for integrals

$$\int e^{i\lambda s^k} \chi(s) ds$$

with $\chi \in C_0^\infty$ (see, *e.g.*, §VIII.1.3 in [180], or §7.7 in [101]).

We need small perturbations of such results. In what follows we set

$$\|g\|_{C^m(I)} := \max_{0\le j\le m} \sup_{x\in I} |g^{(j)}(x)|.$$

Lemma 8.56. *Let $0 < h \le 1$, $I = [-h,h]$, $I^* = [-2h,2h]$ and let $g \in C^2(I^*)$. Suppose that*

$$h \le \frac{1}{10(1+\|g\|_{C^2(I^*)})} \tag{8.57}$$

and let $\eta \in C^1$ be supported in I and satisfy the bounds

$$\|\eta\|_\infty + \|\eta'\|_1 \le A_0, \text{ and } \|\eta'\|_\infty \le A_1. \tag{8.58}$$

Let $k \ge 2$ and define

$$I_\lambda(\eta,x) = \int \eta(s)\exp\Big(i\lambda\Big(\sum_{j=1}^{k-2} x_j s^j + s^k + g(s)s^{k+1}\Big)\Big)ds. \tag{8.59}$$

Let α_k be as in (8.55). Suppose $|x_j| \le \delta\lambda^{\frac{j-k}{k}}$, $j = 1,\dots,k-2$. Then there is an absolute constant C so that, for $\lambda > 2$,

$$|I_\lambda(\eta,x) - \eta(0)\alpha_k\lambda^{-\frac{1}{k}}| \le C[A_0\delta\lambda^{-\frac{1}{k}} + A_1\lambda^{-\frac{2}{k}}(1+\beta_k\log\lambda)];$$

here $\beta_2 = 1$, and $\beta_k = 0$ for $k > 2$.

Proof. We set $u(s) := s(1+sg(s))^{\frac{1}{k}}$; then

$$u'(s) = (1+sg(s))^{-1+\frac{1}{k}}(1+sg(s)+k^{-1}s)$$

and by our assumption on g we quickly verify that

$$\Big(\frac{9}{10}\Big)^{\frac{1}{k}} \le u'(s) \le \Big(\frac{11}{10}\Big)^{\frac{1}{k}}$$

for $-h \le s \le h$. Thus u defines a valid change of variable, with $u(0) = 0$ and $u'(0) = 1$. Denoting the inverse by $s(u)$ we get

$$I_\lambda(\eta,x) = \int \eta_1(u)\exp\Big(i\lambda\Big(\sum_{j=1}^{k-2} x_j s(u)^j + u^k\Big)\Big)du$$

with $\eta_1(u) = \eta(s(u))s'(u)$. Clearly η_1 is supported in $(-2h,2h)$. We observe that

$$\|\eta_1\|_\infty + \|\eta_1'\|_1 \lesssim A_0, \qquad \text{and} \qquad \|\eta_1'\|_\infty \lesssim (A_0h^{-1} + A_1). \tag{8.60}$$

Indeed implicit differentiation and use of the assumption (8.57) reveals that

$$|s''(u)| \lesssim (1+\|g\|_\infty) \lesssim h^{-1}.$$

Taking into account the support properties of η_1 we obtain (8.60).

In order to estimate certain error terms we shall introduce dyadic decompositions. Let $\chi_0 \in C_0^\infty(\mathbb{R})$ so that

$$\chi_0(s) = \begin{cases} 1, & \text{if } |s| \le \frac{1}{4}, \\ 0, & \text{if } |s| \ge \frac{1}{2}, \end{cases} \tag{8.61}$$

and $m \ge 1$, define

$$\chi_m(s) = \chi_0(2^{-m}s) - \chi_0(2^{-m+1}s), \qquad m \ge 1. \tag{8.62}$$

We now split

$$I_\lambda(\eta, x) = \eta_1(0)J_\lambda + \sum_{m\ge 0} E_{\lambda,m} + \sum_{m\ge 0} F_{\lambda,m}(x)$$

where J_λ is defined in (8.54) and

$$E_{\lambda,m} = \int (\eta_1(u) - \eta_1(0))\chi_m(\lambda^{\frac{1}{k}}u)e^{i\lambda u^k}\,du,$$

$$F_{\lambda,m}(x) = \int \eta_1(u)\left(\exp\left(i\lambda\left(\sum_{j=1}^{k-2} x_j s(u)^j\right)\right) - 1\right)\chi_m(\lambda^{\frac{1}{k}}u)e^{i\lambda u^k}\,du.$$

In view of (8.54) the main term in our asymptotics is contributed by $\eta_1(0)J_\lambda$ since $\eta_1(0) = \eta(0)$.

Now we estimate the terms $E_{\lambda,m}$. It is immediate that from an estimate using the support of the amplitude that

$$|E_{\lambda,0}| \le C\|\eta_1'\|_\infty \lambda^{-\frac{2}{k}}.$$

For $m \ge 1$ we integrate by parts once to get

$$E_{\lambda,m} = \frac{i}{k\lambda}\int \frac{d}{du}\big[(\eta_1(u) - \eta_1(0))u^{1-k}\chi_m(\lambda^{\frac{1}{k}}u)\big]e^{i\lambda u^k}\,du$$

and straightforward estimation gives

$$|E_{\lambda,m}| \le C\begin{cases} \|\eta_1'\|_\infty 2^{m(2-k)}\lambda^{-\frac{2}{k}}, & \text{if } 2^m \le \lambda^{\frac{1}{k}} \\ \|\eta_1\|_\infty 2^{m(1-k)}\lambda^{-\frac{1}{k}}, & \text{if } 2^m > \lambda^{\frac{1}{k}}. \end{cases}$$

Thus

$$\sum_m |E_{\lambda,m}| \le C(\|\eta_1\|_\infty \lambda^{-\frac{2}{k}}(1 + \beta_k \log\lambda).$$

We now show that

$$\sum_{m\ge 0} |F_{\lambda,m}(x)| \le C[\|\eta_1\|_\infty + \|\eta'\|_1]\delta\lambda^{-\frac{1}{k}} \tag{8.63}$$

and notice that only terms with $2^m\lambda^{-\frac{1}{k}} \le C$ occur in the sum.

Set

$$\zeta_{\lambda,x}(u) = \left(\exp\left(i\lambda\left(\sum_{j=1}^{k-2} x_j s(u)^j\right)\right) - 1\right).$$

For the term $E_{\lambda,0}(x)$ we simply use the straightforward bound on the support of $\chi_0(\lambda^{\frac{1}{k}}\cdot)$ which is (in view of $|s(u)| \approx |u|$)

$$|\zeta_{\lambda,x}(u)| \leq C\lambda \sum_{j=1}^{k-2} |x_j||\lambda^{-\frac{j}{k}}|$$

and since $|x_j| \leq \delta\lambda^{-\frac{k-j}{k}}$ we get after integrating in u,

$$|F_{\lambda,0}| \lesssim \|\eta_1\|_\infty \delta\lambda^{-\frac{1}{k}}.$$

For $m > 0$ we integrate by parts once and write

$$F_{\lambda,m} = ik^{-1}\lambda^{-1} \int \frac{d}{du}\left[u^{1-k}\chi_m(\lambda^{\frac{1}{k}}u)\eta_1(u)\zeta_{\lambda,x}(u))\right]e^{i\lambda u^k}\, du. \tag{8.64}$$

On the support of $\chi_m(\lambda^{\frac{1}{k}}\cdot)$,

$$|\zeta_{\lambda,x}(u)| \lesssim \lambda \sum_{j=1}^{k-2} \delta\lambda^{-\frac{k-j}{k}}(2^m\lambda^{-\frac{1}{k}})^j \lesssim \delta 2^{m(k-2)}$$

$$\left|\tfrac{d}{du}\left[u^{1-k}\chi_m(\lambda^{\frac{1}{k}}u)\right]\right| \lesssim \lambda 2^{-mk}$$

and also

$$|\zeta'_{\lambda,x}(u)| \lesssim \lambda \sum_{j=1}^{k-2} \delta\lambda^{-\frac{k-j}{k}}(2^m\lambda^{-\frac{1}{k}})^{j-1} \lesssim \delta 2^{m(k-3)}\lambda^{\frac{1}{k}}$$

$$|u^{1-k}\chi_m(\lambda^{\frac{1}{k}}u)| \lesssim 2^{-m(k-1)}\lambda^{\frac{k-1}{k}},$$

and thus we obtain the bound

$$\int \left|\frac{d}{du}\left[u^{1-k}\chi_m(\lambda^{\frac{1}{k}}u)\zeta_{\lambda,x}(u)\eta_1(u))\right]\right| du \lesssim [\|\eta_1\|_\infty + \|\eta_1'\|_1]\delta 2^{-m}\lambda^{\frac{k-1}{k}}.$$

Hence,

$$\sum_{m\geq 0} |F_{\lambda,m}| \lesssim \delta\lambda^{-\frac{1}{k}},$$

which completes the proof of (8.63). □

For the logarithmic lower bounds of $G_4(R)$ we shall need some asymptotics for modifications of Airy functions. Recall that for $t \in \mathbb{R}$ the Airy function is defined by the oscillatory integral

$$Ai(\tau) = \frac{1}{2\pi}\int_{-\infty}^{\infty} \exp\left(i\left(\tfrac{x^3}{3} + \tau x\right)\right) dx$$

and that for $t \to \infty$ we have

$$Ai(-t) = \pi^{-\frac{1}{2}} t^{-\frac{1}{4}} \cos\left(\tfrac{2}{3} t^{\frac{3}{2}} - \tfrac{\pi}{4}\right)\left(1 + O(t^{-\frac{3}{4}})\right). \tag{8.65}$$

This statement can be derived using the method of stationary phase (combining expansions about the two critical points $\pm t^{\frac{1}{2}}$) or complex analysis arguments, *cf.* [66] or [182, p. 330], see also an argument in [90].

Let $g \in C^2([-1,1])$, and let $\varepsilon > 0$ be small, $\varepsilon \lesssim (1 + \|g\|_{C^2})^{-1}$. Let $\eta \in C_0^\infty$ with support in $(-\varepsilon, \varepsilon)$, so that $\eta(s) = 1$ for $|s| \le \frac{\varepsilon}{2}$.

Lemma 8.66. *Define*

$$J(\lambda, \vartheta) = \int e^{i\lambda\left(\frac{s^3}{3} - \vartheta s\right)} e^{i\lambda g(s) s^4} \eta(s) ds. \tag{8.67}$$

Then, for $0 < \vartheta < \frac{\varepsilon^2}{2}$ and $\lambda > \varepsilon^{-1}$,

$$\begin{aligned} J(\lambda, \vartheta) &= \lambda^{-\frac{1}{3}} Ai(-\lambda^{\frac{2}{3}} \vartheta) + E_1(\lambda, \vartheta) \\ &= \pi^{-\frac{1}{2}} \lambda^{-\frac{1}{2}} \vartheta^{-\frac{1}{4}} \cos\left(\tfrac{2}{3} \lambda \vartheta^{\frac{3}{2}} - \tfrac{\pi}{4}\right) + E_2(\lambda, \vartheta) \end{aligned} \tag{8.68}$$

where, for $i = 1, 2$,

$$|E_i(\lambda, \vartheta)| \lesssim C_\varepsilon \left[\lambda^{-1} \vartheta^{-1} + \min\left\{\lambda \vartheta^{\frac{5}{2}}, \vartheta^{\frac{1}{2}}\right\}\right]. \tag{8.69}$$

Proof. We split

$$J(\lambda, \vartheta) = \sum_{i=1}^4 J_i(\lambda, \vartheta) := \sum_{i=1}^4 \int e^{i\lambda\left(\frac{s^3}{3} - \vartheta s\right)} \zeta_i(s)\, ds$$

where

$$\begin{aligned} &\zeta_1(s) = 1, \qquad \zeta_2(s) = (\eta(s) - 1), \\ &\zeta_3(s) = \eta(s)(e^{i\lambda g(s) s^4} - 1)\eta(C^{-1} \vartheta^{-\frac{1}{2}} s), \\ &\zeta_4(s) = \eta(s)(e^{i\lambda g(s) s^4} - 1)(1 - \eta(C^{-1} \vartheta^{-\frac{1}{2}} s)) \end{aligned}$$

where $C \ge \varepsilon^{-1}$. By a scaling we see that

$$J_1(\lambda, \vartheta) = \lambda^{-\frac{1}{3}} Ai(-\lambda^{\frac{2}{3}} \vartheta)$$

and we prove upper bounds for the error terms J_i, $i = 2, 3, 4$. Let

$$\Phi(s) = -\vartheta s + \frac{s^3}{3},$$

then $\Phi'(s) = -\vartheta + s^2$ and in the support of ζ_2 we have $|\Phi'(s)| \ge c\varepsilon$. Thus by an integration by parts $J_2(\lambda, \vartheta) = O(\lambda^{-1})$. Note that ζ_3 is bounded and that also

$|\zeta_3(s)| \lesssim \lambda|\vartheta|^2$. We integrate over the support of ζ_3 which is of length $O(\sqrt{b})$ and obtain $J_3(\lambda,\vartheta) = O(\min\{\vartheta^{\frac12}, \lambda\vartheta^{\frac52}\})$. To estimate $J_4(\lambda,\vartheta)$ we argue by van der Corput's Lemma, for the phases Φ and its perturbation $\Psi(s) := \Phi(s) + s^4 g(s)$. Thus we split

$$J_4(\lambda,\vartheta) = \sum_m \sum_\pm J_{4,m,\pm}(\lambda,\vartheta)$$

where we have set

$$J_{4,m,\pm}(\lambda,\vartheta) = \int e^{i\lambda\Psi(s)} \rho_{m,\pm}(s)\,ds - \int e^{i\lambda\Phi(s)} \rho_{m,\pm}(s)\,ds;$$

here

$$\rho_{m,+}(s) = \chi_{(0,\infty)}\eta(s)(1-\eta(C^{-1}\vartheta^{-\frac12}s))\chi_m(C^{-1}\vartheta^{-\frac12}s),$$

χ_m is as in (8.62) and $2^m\vartheta^{\frac12} \lesssim \varepsilon$ (in view of the condition on η). Let $\rho_{m,-}$ is analogously defined, with support on $(-\infty,0)$.

We argue as in the proof of Lemma 8.56. Note that now $|\Phi'(s)| \approx 2^{2m}\vartheta$,

$$\frac{\partial}{\partial s}\left(g(s)s^4\right) = O(2^{3m}\vartheta^{\frac32})$$

and since $2^m\vartheta^{\frac12} \lesssim \varepsilon$, we also have $|\Psi'(s)| \approx 2^{2m}\vartheta$. Moreover, observe that $\Phi''(s) = 2s + O(s^2)$ so that van der Corput's Lemma can be applied to the two integrals defining $J_{4,m,\pm}(\lambda,\vartheta)$. We obtain $J_{4,m,\pm}(\lambda,\vartheta) = O(\lambda^{-1}\vartheta^{-1}2^{-2m})$.

Finally, by (8.65) and (8.68), the difference of E_1 and E_2 is $O(\lambda^{-1}\vartheta^{-1})$. This concludes the proof. □

8.6 Lower bounds

For $w \in \mathbb{R}^d$ (usually restricted to the unit sphere), define

$$F_R(w) = \int \chi(t)e^{iR\langle\gamma(t),w\rangle}\,dt. \tag{8.70}$$

The following result establishes inequality (8.13) of Theorem 8.12.

Proposition 8.71. *Suppose that for some $t_0 \in I$ the vectors $\gamma'(t_0),\dots,\gamma^{(k)}(t_0)$ are linearly independent. Then $\chi \in C_0^\infty$ in (8.1) can be chosen so that, for sufficiently large R,*

$$\|F_R\|_{L^q(S^{d-1})} \geq CR^{-\frac1k - \frac{k^2-k-2}{2kq}}. \tag{8.72}$$

Proof. We may assume $t_0 = 0$. By a scaling and rotation we may assume that $\gamma^{(k)}(0) = e_k$. We shall then show the lower bound $|F_R(\omega)| \geq c_0 R^{-\frac1k}$ for a neighborhood of e_k which is of measure $\approx R^{-\frac{k^2-k-2}{2k}}$. Now let A_k be an invertible linear transformation which maps e_k to itself, and for $j = 1, \dots, k-1$ maps $\gamma^{(j)}(0)$ to e_j, $j = 1, \dots, k$. Then the map

$$\omega \to \frac{(A_k^*)^{-1}\omega}{|(A_k^*)^{-1}\omega|}$$

defines a diffeomorphism from a spherical neighborhood of e_k to a spherical neighborhood of e_k. Thus we may assume for what follows that $\gamma : [-1, 1] \to \mathbb{R}^d$ satisfies

$$\gamma^{(j)}(0) = e_j, \qquad j = 1, \dots, k. \tag{8.73}$$

We may also assume that the cutoff function χ is supported in a small open interval $(-\varepsilon, \varepsilon)$ so that $\chi(0) = 1$.

As we have $\langle e_k, \gamma^{(k-1)}(0)\rangle = 0$ and $\langle e_k, \gamma^k(0)\rangle = 1$ we can use the implicit function theorem to find a neighborhood $\mathcal{W}_k$ of e_k and an interval $\mathcal{I}_k = (-\varepsilon_k, \varepsilon_k)$ containing 0 so that for all $w \in \mathcal{W}_k$ the equation $\langle w, \gamma^{(k-1)}(t)\rangle = 0$ has a unique solution $\widetilde{t}_k(w) \in \mathcal{I}_k$. This solution is also homogeneous of degree 0, *i.e.*, $\widetilde{t}_k(sw) = \widetilde{t}_k(w)$ for s near 1), and we have $\widetilde{t}_k(e_k) = 0$.

Lemma 8.74. *There is $\varepsilon_0 > 0$, $R_0 > 1$, and $c > 0$ so that for all positive $\varepsilon < \varepsilon_0$ and all $R > R_0$ the following holds. Let*

$$\begin{aligned} U_{k,\varepsilon}(R) = \{\omega \in S^{d-1} : |\omega - e_k| \le \varepsilon \text{ and} \\ |\langle \omega, \gamma^{(j)}(\widetilde{t}_k(\omega))\rangle| \le \varepsilon R^{\frac{j-k}{k}}, \text{ for } j = 1, \dots, k-2.\} \end{aligned}$$

Then the spherical measure of $U_{k,\delta}(R)$ is at least $c\varepsilon^{d-1}R^{-\frac{k^2-k-2}{2k}}$.

Proof. In a neighborhood of e_k we parametrize the sphere by

$$\omega(y) = (y_1, \dots, y_{k-1}, \sqrt{1-|y|^2}, y_k, \dots, y_{d-1}).$$

We introduce new coordinates $z_1, \dots, z_{d-1}$ setting

$$z_j = \mathfrak{z}_j(y) = \begin{cases} \langle \omega(y), \gamma^{(j)}(\widetilde{t}_k(\omega(y)))\rangle, & j = 1, \dots, k-2, \\ y_j, & j = k-1, \dots, d-1. \end{cases} \tag{8.75}$$

Then it is easy to see that $\mathfrak{z}$ defines a diffeomorphism between small neighborhoods of the origin in $\mathbb{R}^{d-1}$; indeed the derivative at the origin is the identity map.

The spherical measure of $U_{k,\varepsilon}(R)$ is comparable to the measure of the set of $z \in \mathbb{R}^{d-1}$ satisfying $|z_j| \le \varepsilon R^{\frac{j-k}{k}}$, for $j = 1, \dots, k-2$, and $|z_j| \le \varepsilon$ for $k-1 \le j \le d-1$, and this set has measure $\approx \varepsilon^{d-1}R^{-\frac{k^2-k-2}{2k}}$. □

We now verify that for sufficiently small ε and sufficiently large R,

$$|F_R(\omega)| \ge c_0 R^{-\frac{1}{k}}, \quad \omega \in U_{k,\varepsilon}(R), \tag{8.76}$$

with some positive constant c_0; by Lemma 8.74, this of course implies the bound (8.72). To see (8.76) we set

$$a_j(\omega) = \langle \omega, \gamma^{(j)}(\widetilde{t}_k(\omega))\rangle, \tag{8.77}$$

$s = t - \widetilde{t}_k(\omega)$ and expand

$$\langle \omega, \gamma(t)\rangle - \langle \omega, \gamma(\widetilde{t}_k(\omega))\rangle = \sum_{j=1}^{k-2} a_j(\omega)\frac{s^j}{j!} + a_k(\omega)\frac{s^k}{k!} + \mathcal{E}_k(\omega, s)s^{k+1}, \tag{8.78}$$

with

$$\mathcal{E}_k(\omega, s) = \int_{\sigma=0}^{1} \tfrac{(1-\sigma)^k}{k!}\langle \omega, \gamma^{(k+1)}(\widetilde{t}_k(\omega) + \sigma s)\rangle\, d\sigma.$$

If ε is sufficiently small then we can apply Lemma 8.56 with $\omega \in U_{k,\varepsilon}(R)$, and the choice $\lambda = \frac{R}{k!}\langle \omega, \gamma^{(k)}(\widetilde{t}_k(\omega))\rangle$, and the lower bound (8.76) follows. □

We now formulate bounds for $q \geq \frac{k^2+k+2}{2}$ for the case that $\gamma^{(k+1)} \equiv 0$ for some $k < d$; this of course implies that the curve lies in a k-dimensional affine subspace.

Proposition 8.79. *Suppose that γ is a polynomial curve with $\gamma^{(k+1)} \equiv 0$ and suppose that for some $t_0 \in I$ the vectors $\gamma'(t_0), \dots, \gamma^{(k)}(t_0)$ are linearly independent. Then χ in* (8.1) *can be chosen so that for sufficiently large R,*

$$\|F_R\|_{L^q(S^{d-1})} \geq C \begin{cases} R^{-\frac{k}{q}}[\log R]^{\frac{1}{q}}, & q = \frac{k^2+k+2}{2}, \\ R^{-\frac{k}{q}}, & q > \frac{k^2+k+2}{2}. \end{cases} \tag{8.80}$$

Proof. We first note that the assumption $\gamma^{(k+1)} \equiv 0$ implies that the curve is polynomial and for any fixed t_0 it stays in the affine subspace through $\gamma(t_0)$ which is generated by $\gamma^{(j)}(t_0)$, $j = 1, \dots, k$. We shall prove a lower bound for $\widehat{\mu}$ in a neighborhood of a vector $e \in S^{d-1}$ where e is orthogonal to the vectors $\gamma^{(j)}(t_0)$. After a rotation we may assume that

$$\gamma(t) = (\gamma_1(t), \dots, \gamma_k(t), 0, \dots, 0).$$

For $\omega \in S^{d-1}$, we split accordingly $\omega = (\omega', \omega'')$ with small $\omega' \in \mathbb{R}^k$, namely

$$|\omega'| \approx 2^{-l},$$

where $1 \lesssim R2^{-l} \lesssim R$. As before, we solve the first degree equation

$$\langle \gamma^{(k-1)}(t), \omega\rangle = 0$$

(observe that this is actually independent of ω'') with $t = \widetilde{t}_k(\omega')$; now $\widetilde{t}_k$ is homogeneous of degree 0 as a function on $\mathbb{R}^k$. Then

$$e^{-i\langle \omega, \gamma(\widetilde{t}_k(\omega))\rangle} F_R(\omega)$$
$$= \int \chi(\widetilde{t}_k(\omega') + s) \exp\left(\sum_{j=1}^{k-2}\langle \omega, \gamma^{(j)}(\widetilde{t}_k(\omega'))\rangle \tfrac{s^j}{j!} + \langle \omega, \gamma^{(k)}(\widetilde{t}_k(\omega'))\rangle \tfrac{s^k}{k!}\right) ds.$$

If $k \geq 3$, let $V_{k,l}(R)$ be the subset of the unit sphere in $\mathbb{R}^k$ which consists of those $\theta \in S^{k-1}$ which satisfy the conditions

$$|\langle \gamma^{(\nu)}(\widetilde{t}_k(\theta)), \theta\rangle| \leq \varepsilon (R2^{-l})^{\frac{k-\nu}{k}}, \quad \nu = 1, \dots, k-2.$$

Observe that the spherical measure of $V_{k,l}(R)$ (as a subset of S^{k-1}) is

$$(R2^{-l})^{-\frac{k^2-k-2}{2k}},$$

by Lemma 8.74. Now, if $k = 2$, define

$$U_{2,l}(R) := \{\omega = (\omega', \omega'') \in S^{d-1} : |\omega - e_3| \leq \delta, 2^{-l} \leq |\omega'| < 2^{-l+1}\}.$$

If $3 \leq k < d$ let

$$U_{k,l}(R) := \{\omega \in S^{d-1} : |\omega - e_{k+1}| \leq \delta, \\ 2^{-l} \leq |\omega'| < 2^{-l+1}, \tfrac{\omega'}{|\omega'|} \in V_{k,l}(R)\}.$$

We need a lower bound for the spherical measure (on S^{d-1}) of $U_{k,l}(R)$ and using polar coordinates in $\mathbb{R}^k$ we see that it is at least

$$c\varepsilon^{d-1} 2^{-lk} (R2^{-l})^{-\frac{k^2-k-2}{2k}}.$$

If ε is small we obtain a lower bound $c(R2^{-l})^{-\frac{1}{k}}$ on this set; this follows from Lemma 8.56 with $\lambda \approx R2^{-l}$. Thus

$$\int_{U_{k,l}(R)} |F_R(\omega)|^q d\sigma(\omega) \geq c_\varepsilon (R2^{-l})^{-\frac{q}{k}} 2^{-lk} (R2^{-l})^{-\frac{k^2-k-2}{2k}} \\ = c_\varepsilon R^{-\frac{q}{k} - \frac{k^2-k-2}{2k}} 2^{l\left(\frac{q}{k} - \frac{k^2+k+2}{2k}\right)}.$$

As the sets $U_{k,l}(R)$ are disjoint in l we may now sum in l for $CR^{-1} \leq 2^{-l} \leq c$ for a large C and a small c. Then we obtain that

$$\sum_l \int_{U_{k,l}(R)} |F_R(\omega)|^q d\sigma(\omega)$$

is bounded below by $cR^{-\frac{q}{k} - \frac{k^2-k-2}{2kq}}$, if $q < \frac{k^2+k+2}{2}$; this yields the bound that was already proved in Proposition 8.71. If $q > \frac{k^2+k+2}{2}$ then we get the lower bound cR^{-k} and for the exponent $q = \frac{k^2+k+2}{2}$ we obtain the lower bound $cR^{-k} \log R$. This yields (8.80). □

Proposition 8.81. *Suppose that $3 \leq k \leq d$ and that for some $t_0 \in I$ the vectors $\gamma'(t_0), \dots, \gamma^{(k)}(t_0)$ are linearly independent. Then $\chi \in C_0^\infty$ in (8.1) can be chosen so that for sufficiently large R,*

$$\|F_R\|_{L^q(S^{d-1})} \geq CR^{-\frac{k-1}{q}} [\log R]^{\frac{1}{q}}, \quad \text{if } q = q_{k-1} = \tfrac{k^2-k+2}{2}. \tag{8.82}$$

Proof. We start with the same reductions as in the proof of Proposition 8.71, namely we may assume $t_0 = 0$ and $\gamma^{(j)}(0) = e_j$ for $j = 1, \dots, k$; we shall then derive lower bounds for $F_R(\omega)$ for ω near e_k. As before denote by $\widetilde{t_k}(\omega)$ the solution t of $\langle \gamma^{(k-1)}(t), \omega \rangle = 0$, for ω near e_k. We may use the expansion (8.78). Define the polynomial approximation

$$P_k(s, \omega) = P_k(s) = \sum_{j=1}^{k-2} a_j(\omega) \frac{s^j}{j!} + a_k(\omega) \frac{s^k}{k!}.$$

Note that $a_k(\omega)$ is near 1 if ω is near e_k. In what follows we shall only consider those ω with

$$a_{k-2}(\omega) < 0.$$

In our analysis we need to distinguish between the cases $k = 3$ and $k > 3$.

The case $k = 3$. We let for small δ

$$\mathcal{U}_{R,j} = \{\omega \in S^{d-1} : |\omega - e_3| \le \delta,\ -2^{j+1} R^{-\frac{2}{3}} \le a_1(\omega) \le -2^j R^{-\frac{2}{3}}\}.$$

We wish to use the asymptotics of Lemma 8.66, with the parameters

$$\vartheta = \vartheta(\omega) = \frac{-2a_1(\omega)}{a_3(\omega)}$$

and $\lambda = \frac{R}{2} a_3(\omega)$ ($\approx R$) to derive a lower bound on a portion of $\mathcal{U}_{R,j}$ whenever $\lambda^{-\frac{2}{3}} \lesssim \vartheta(\omega) \lesssim \lambda^{-\frac{6}{11}}$; *i.e.*,

$$\delta^{-1} \le 2^j \le \delta \lambda^{\frac{4}{33}} \tag{8.83}$$

where δ is small (but independent of large λ).

The range (8.83) is chosen so that the error terms in (8.69) (with $\lambda \approx R$) are $\lesssim R^{-\frac{1}{2}} \vartheta^{-\frac{1}{4}}$ if δ is small; indeed the term $\lambda^{-1}\vartheta^{-1}$ is controlled by $C\delta^{\frac{3}{4}} \lambda^{-\frac{1}{2}} \vartheta^{-\frac{1}{4}}$ in view of the first inequality in (8.83) and the term $\lambda \vartheta^{\frac{5}{2}}$ is bounded by $C\delta^{\frac{11}{4}} \lambda^{-\frac{1}{2}} \vartheta^{-\frac{1}{4}}$ because of the second restriction. Since the main term in (8.68) can be written as

$$\left(\frac{2}{\pi}\right)^{\frac{1}{2}} R^{-\frac{1}{2}} a_3(\omega)^{-\frac{1}{2}} \vartheta(\omega)^{-\frac{1}{4}} \cos\left(\tfrac{1}{3} R a_3(\omega) \vartheta(\omega)^{\frac{3}{2}} - \tfrac{\pi}{4}\right),$$

it dominates the error terms in the range (8.83), provided that we stay away from the zeroes of the cosine term. To achieve the necessary further localization we let, for positive integers n,

$$\mathcal{U}_{R,j,n} = \left\{\omega \in \mathcal{U}_{R,j} : \left|\tfrac{1}{3} R a_3(\omega) \vartheta(\omega)^{\frac{3}{2}} - \tfrac{\pi}{4} - \pi n\right| < \tfrac{\pi}{4}\right\}.$$

Let j be in the range (8.83). We use

$$|b^{\frac{3}{2}} - a^{\frac{3}{2}}| \approx \left(\sqrt{a} + \sqrt{b}\right) |b - a|$$

for $0 < b, a \lesssim 1$. Since $\vartheta(\omega)$ can be used as one of the coordinates on the unit sphere we see that the spherical measure of $\mathcal{U}_{R,j,n}$ is $\gtrsim \delta^2 R^{-\frac{2}{3}} 2^{-\frac{j}{2}}$ for the about $2^{\frac{3j}{2}}$ values of n for which $n \approx 2^{\frac{3j}{2}}$, and on those disjoint sets $\mathcal{U}_{R,j,n}$ the value of $F_R(\omega)$ is $\geq cR^{-\frac{1}{3}} 2^{-\frac{j}{4}}$.

This implies that, for j as in (8.83),

$$\text{meas}\left(\left\{\omega \in \mathcal{U}_{R,j} : |F_R(\omega)| \geq c_\delta R^{-\frac{1}{3}} 2^{-\frac{j}{4}}\right\}\right) \geq c'_\delta 2^j R^{-\frac{2}{3}},$$

and thus

$$\int_{\mathcal{U}_{R,j}} |F_R(\omega)|^4 d\sigma(\omega) \gtrsim R^{-2}.$$

Since the sets $\mathcal{U}_{R,j}$ are disjoint we may sum in j over the range (8.83) and obtain the lower bound

$$\|F_R\|_4 \gtrsim R^{-\frac{1}{2}} (\log R)^{\frac{1}{4}},$$

with an implicit constant depending on δ.

The case $k > 3$. We try to follow in spirit the proof of the case for $k = 3$. Notice that

$$P_k^{(k-2)}(s) = a_{k-2}(\omega) + \frac{1}{2} a_k(\omega) s^2$$

has then two real roots, one of them being

$$s_1(\omega) = \left(\frac{-2a_{k-2}(\omega)}{a_k(\omega)}\right)^{\frac{1}{2}},$$

the other one $s_2 = -s_1$. The idea is now to use, for suitable ω, an asymptotic expansion for the part where s is close to s_1, and, unlike in the case $k = 3$, we shall now be able to neglect the contribution of the terms where s is near s_2. To achieve this we define, for $j = 1, \dots, k-3$,

$$\widetilde{a}_j(\omega) = P_k^{(j)}(s_1(\omega)) = a_j(\omega) \quad + \sum_{\substack{1 \leq \nu \leq k-2-j \\ \text{or } \nu = k-j}} \frac{a_{j+\nu}(\omega)}{\nu!} \left(\frac{-2a_{k-2}(\omega)}{a_k(\omega)}\right)^{\frac{\nu}{2}}. \tag{8.84}$$

We further restrict consideration to ω chosen in sets

$$\begin{aligned} \mathcal{V}_{k,j}(\delta) = \big\{\omega \in S^{d-1} : -2^{j+1} R^{-\frac{2}{k}} \\ < a_{k-2}(\omega) < -2^j R^{-\frac{2}{k}}, |e_k - \omega| \leq \delta, \\ |\widetilde{a}_\nu(\omega)| \leq \delta |a_{k-2}(\omega)|^{\frac{\nu}{2k-2}} R^{-\frac{k-\nu-1}{k-1}}, 1 \leq \nu \leq k-3\big\}. \end{aligned} \tag{8.85}$$

We shall see that if we choose ω from one of the sets $\mathcal{V}_{k,j}(\delta)$ with small δ, and j not too large, then the main contribution of the oscillatory integral comes from the part where $|s - s_1(\omega)| \leq \frac{1}{2} s_1(\omega)$. We shall reduce to an application of Lemma

8.56 to derive a lower bound for that part. For the remaining parts we shall derive smaller upper bounds using van der Corput's Lemma.

For notational convenience we abbreviate

$$b := -a_{k-2}(\omega), \quad \widetilde{a}_\nu := \widetilde{a}_\nu(\omega), \quad s_1 := s_1(\omega), \quad \widetilde{t}_k := \widetilde{t}_k(\omega).$$

We now split

$$e^{-i\langle\omega,\gamma(\widetilde{t}_k(\omega))\rangle} F_R(\omega) = I_R(\omega) + E_R(\omega), \tag{8.86}$$

where

$$I_R(\omega) = \int \chi(\widetilde{t}_k + s)\chi_0\left(20\frac{s - s_1}{s_1}\right) \exp\left(iR[P_k(s) + s^{k+1}\mathcal{E}_{k+1}(s,\omega)]\right) ds.$$

Here χ_0 is as in (8.61) and thus the integrand is supported where $|s - s_1| \le s_{\frac{1}{40}}$.

Notice that $P_k^{(k-1)}(s_1) = s_1 a_k(\omega)$ and $P_k^{(k)}(s) \equiv a_k(\omega)$. Let

$$Q_{k-1}(s) = \sum_{\nu=1}^{k-3} \widetilde{a}_\nu \frac{(s - s_1)^\nu}{\nu!} + a_k s_1 \frac{(s - s_1)^{(k-1)}}{(k-1)!};$$

then

$$P_k(s) - P_k(s_1) = Q_{k-1}(s) + \frac{1}{k!} a_k (s - s_1)^k.$$

Thus we can write

$$I_R(\omega) = \int \eta(s) e^{iR(Q_{k-1}(s) + \frac{1}{k!} a_k (s - s_1)^k)} ds$$

with

$$\eta(s) = \chi(\widetilde{t}_k + s)\chi_0(10 s_1^{-1}(s - s_1)) \exp(iRs^{k+1}\mathcal{E}_{k+1}(s,\omega)).$$

Note that by (8.61) the function η is supported where $20 s_1^{-1}|s - s_1| \le \frac{1}{2}$, *i.e.*, in $[s_1 - h, s_1 + h]$ with $h = s_{\frac{1}{40}}$. Clearly $\|\eta\|_\infty = O(1)$, and since $s_1 \approx \sqrt{b}$ it is straightforward to check that

$$\|\eta'\|_\infty + b^{-\frac{1}{2}}\|\eta'\|_1 \lesssim (1 + b^{-\frac{1}{2}} + Rb^{\frac{k}{2}}), \tag{8.87}$$

thus also

$$\|\eta\|_\infty + \|\eta'\|_1 \lesssim 1 \text{ if } b \le R^{-\frac{2}{k+1}}. \tag{8.88}$$

Moreover, if $g(s) = \frac{1}{ks_1}$ then we can write

$$RQ_{k-1}(s) + a_k \frac{(s - s_1)^k}{k!}$$
$$= \frac{Ra_k s_1}{(k-1)!} \left(\sum_{\nu=1}^{k-3} x_\nu (s - s_1)^\nu + (s - s_1)^{k-1} + (s - s_1)^k g(s - s_1) \right)$$

where $|x_\nu| \lesssim b^{-\frac{1}{2}}|\widetilde{a}_\nu|$. The conditions

$$|\widetilde{a}_\nu| \le \delta b^{\frac{\nu}{2k-2}} R^{-\frac{k-\nu-1}{k-1}} \qquad \text{imply that} \qquad |x_\nu| \lesssim \delta(Rb^{\frac{1}{2}})^{-\frac{k-\nu-1}{k-1}}.$$

We of course have

$$\|g\|_{C^2([-h,h])} \le s_1^{-1} \qquad \text{on} \qquad I^* = \big[-s_{\frac{1}{10}}, s_{\frac{1}{10}}\big];$$

thus

$$h = s_{\frac{1}{40}} \le 10^{-1}(1 + \|g\|_{C^2})^{-1}.$$

Changing variables $\tilde{s} = s - s_1$ puts us in the position to apply Lemma 8.56 for perturbations of the phase $\tilde{s} \mapsto \lambda \tilde{s}^{k-1}$, with

$$\lambda := R|a_k|s_1 = R\sqrt{2a_k b} \approx Rb^{\frac{1}{2}},$$

and we have the bounds $A_0 \le C$ (if $b \le R^{-\frac{2}{k+1}}$ and $A_1 \le (1 + Rb^{\frac{k}{2}})$ for the parameters in Lemma 8.56. We thus obtain (*cf.* (8.55))

$$\begin{aligned} &\big|I_R(\omega) - \alpha_{k-1}\chi(s_1(\omega))(R\sqrt{2a_k b})^{-\frac{1}{k-1}}\big| \\ &\quad \lesssim \delta(Rb^{\frac{1}{2}})^{-\frac{1}{k-1}} + b^{-\frac{1}{2}}(Rb^{\frac{1}{2}})^{-\frac{2}{k-1}}\log(Rb^{\frac{1}{2}}), \end{aligned} \tag{8.89}$$

provided that $b \le R^{-\frac{2}{k+1}} \lesssim 1$. We wish to use this lower bound on the sets $\mathcal{V}_{k,j}(\delta)$. In order to efficiently apply (8.89) we shall choose j so that

$$R^{-\tau_1+\frac{2}{k}} \le 2^j \le R^{-\tau_2+\frac{2}{k}}, \tag{8.90}$$

with τ_1, τ_2 satisfying

$$\frac{2}{k} > \tau_1 > \tau_2 > \frac{2}{k+1},$$

so that the main term in (8.89) dominates the error terms.

We now need to bound from below the measure of the set $\mathcal{V}_{k,j}(\delta)$. We use the coordinates (8.75) on the sphere in a neighborhood of e_k. In view of the linear independence of $\gamma'', \dots, \gamma^{(k-1)}$ we can use the functions $a_j(\mathfrak{z}(y))$, $j \in \{1, \dots, k-2\}$, *cf.* (8.77), as a set of partial coordinates.

We may also change coordinates

$$(a_1, \dots, a_{k-3}, a_{k-2}) \mapsto (\widetilde{a}_1, \dots, \widetilde{a}_{k-3}, a_{k-2}),$$

with $a_{k-2} \equiv -b$; here we use the shear structure of the (nonsmooth) change of variable (8.84). Thus, as in Lemma 8.74, we obtain a lower bound for the spherical measure of $\mathcal{V}_{k,j}(\delta)$, namely

$$\begin{aligned} |\mathcal{V}_{k,j}(\delta)| &\ge c\delta^{d-2}2^j R^{-\frac{2}{k}} \prod_{\nu=1}^{k-3} \Big((2^j R^{-\frac{2}{k}})^{\frac{\nu}{2(k-1)}} R^{-\frac{k-\nu-1}{k-1}} \Big) \\ &= c\delta^{d-2}2^j R^{-\frac{2}{k}} (2^j R^{-\frac{2}{k}})^{\frac{(k-3)(k-2)}{4(k-1)}} R^{\frac{(k-3)(k-2)}{2(k-1)} - (k-3)} \\ &= c\delta^{d-2} 2^{j\frac{k^2-k+2}{4(k-1)}} R^{-\frac{k^2-k-2}{2k}} \end{aligned}$$

after a little arithmetic. Thus

$$\left|\mathcal{V}_{k,j}(\delta)\right| \geq c\delta^{d-2} 2^{\frac{jq_{k-1}}{2k-2}} R^{\frac{1}{k}-\frac{k-1}{2}}. \tag{8.91}$$

Now if δ is chosen small and then fixed, and R is chosen large, then (8.89) implies the lower bound

$$\begin{aligned} |I_R(\omega)| &\geq c_\delta \big(R\sqrt{2^j R^{-\frac{2}{k}}}\big)^{-\frac{1}{(k-1)}} \\ &= c_\delta 2^{-\frac{j}{(2k-2)}} R^{-\frac{1}{k}}, \quad \omega \in \mathcal{V}_{k,j}(\delta), \end{aligned} \tag{8.92}$$

provided that $R^{-\tau_1+\frac{2}{k}} \leq 2^j \leq R^{-\tau_2+\frac{2}{k}}$. We shall verify that for $j \geq 0$,

$$|E_R(\omega)| \lesssim R^{-\frac{1}{k}}\big(2^{-\frac{j}{k-2}} + 2^{-\frac{3j}{2k-6}}\big), \quad \omega \in \mathcal{V}_{k,j}(\delta), \tag{8.93}$$

and from (8.92) and (8.93) it follows that

$$|F_R(\omega)| \geq c_\delta 2^{-\frac{j}{2k-2}} R^{-\frac{1}{k}}, \quad \omega \in \mathcal{V}_{k,j}(\delta)$$

if $R^{-\tau_1+\frac{2}{k}} \leq 2^j \leq R^{-\tau_2+\frac{2}{k}}$. By (8.91) this implies for the same range a lower bound which is independent of j,

$$\int_{\mathcal{V}_{k,j}(\delta)} |F_R(\omega)|^{q_{k-1}} d\omega \geq c_\delta R^{-\frac{q_{k-1}}{k} - \frac{k^2-k-2}{2k}} = c_\delta R^{-(k-1)}.$$

We sum in j, $R^{-\tau_1+\frac{2}{k}} \leq 2^j \leq R^{-\tau_2+\frac{2}{k}}$; this yields, for large R,

$$\left(\int_{\cup_j \mathcal{V}_{k,j}(\delta)} |F_R(\omega)|^{q_{k-1}} d\omega\right)^{\frac{1}{q_{k-1}}} \geq c'_\delta R^{-\frac{k-1}{q_{k-1}}} \big(\log R\big)^{\frac{1}{q_{k-1}}}$$

which is the desired bound.

It remains to prove the upper bounds (8.93) for the error term E_R. It is given by

$$E_R(\omega) = e^{iRP_k(0)} \int \chi(\widetilde{t}_k + s)\left(1 - \chi_0\Big(20\frac{s-s_1}{s_1}\Big)\right) e^{iR\phi(s)} ds$$

where

$$\phi(s) = P_k(s) - P_k(0) + s^{k+1}\mathcal{E}_{k+1}(s).$$

We use a simple application of van der Corput's Lemma. Write ϕ as

$$\phi(s) = Q_{k-1}(s) + \frac{1}{k!}a_k(s-s_1)^k + s^{k+1}\mathcal{E}_{k+1}(s).$$

and observe

$$\phi^{(k-2)}(s) = \frac{1}{2}a_k(s-s_1)(s+s_1) + O(s^3),$$
$$\phi^{(k-3)}(s) = \widetilde{a}_{k-3} + a_k\frac{(s-s_1)^2}{2}\left(\frac{2s_1}{3}+\frac{s}{3}\right) + O(s^4).$$

The integrand of the integral defining E_R is supported where $|s-s_1| \geq \frac{s_1}{80}$, and $|s-s_1| \leq c$ for small c. We see that

$$|\phi^{(k-2)}(s)| \geq c_0 b$$

if in addition $|s+s_1| \geq \frac{s_1}{10}$.

If $|s+s_1| \leq \frac{s_1}{10}$, this lower bound breaks down; however, we have then

$$|\phi^{(k-3)}(s)| \geq cb^{\frac{3}{2}} - |\widetilde{a}_{k-3}(\omega)|.$$

Now on $\mathcal{V}_{k,j}(\delta)$ we have the restriction

$$|\widetilde{a}_{k-3}(\omega)| \leq \delta b^{\frac{k-3}{2(k-1)}} R^{-\frac{2}{k-1}} \leq \delta b^{\frac{3}{2}}$$

where the last inequality is equivalent to the imposed condition $b \geq R^{-\frac{2}{k}}$ (which holds when $j \geq 2$). Thus if δ is small we have

$$|\phi^{(k-3)}(s)| \approx b^{\frac{3}{2}}$$

if $|s+s_1| \leq s_{\frac{1}{10}}$.

We now split the integral into three parts (using appropriate adapted cutoff functions), namely where (i) $|s+s_1| \leq s_{\frac{1}{10}}$, or (ii) $s+s_1 \geq s_{\frac{1}{10}}$, or (iii) $s+s_1 \leq -s_{\frac{1}{10}}$. For parts (ii) and (iii) we can use van der Corput's Lemma with $k-2$ derivatives and see that the corresponding integrals are bounded by $C(Rb)^{-\frac{1}{k-2}}$. Similarly for part (i), if $k > 4$ we can use van der Corput's Lemma with $(k-3)$ derivatives to see that the corresponding integral is bounded by $C(Rb^{\frac{3}{2}})^{-\frac{1}{k-3}}$. The case $k=4$ requires a slightly different argument (as we do not necessarily have adequate monotonicity properties on ϕ'), however in the region (i) we now have

$$\phi''(s) = O(b), \quad |\phi'(s)| \gtrsim b^{\frac{3}{2}}$$

and integrating by parts once gives the required bound $O\left(\frac{1}{Rb^{\frac{3}{2}}}\right)$ also in this case. Since $b \approx R^{-\frac{2}{k}}2^j$, the upper bound (8.93) follows. □

Epilogue

Now that the book has ended, the time has come to take stock of what we have done. The chapters above already contain much information about possible directions for further research on the level of details. The introduction places the techniques and ideas of this book into context of modern research trends and explains some of the connections between them. To wind things down, we describe using broad strokes some of the fundamental gaps in the current state of knowledge as it pertains to the material of this book.

The first part of the book examines the decay rate of Fourier transforms of functions based on their analytic properties. The method of stationary phase and related techniques are reviewed and then more delicate assumptions are considered, which leads to the study of functions of bounded variation. Perhaps the most interesting examples are provided by the connections between these problems and the various properties of the Hilbert transform and the real Hardy spaces. Similar results are also presented in the higher-dimensional case with the radial case providing a natural transition point. It is the higher-dimensional setup where our state of knowledge is particularly incomplete and this should serve as a good starting point for a variety of future investigations.

The second half of this book is built heavily around the concept of average decay of the Fourier transform. While the relationship between curvature and the decay of the Fourier transform is fairly well understood, the role of smoothness remains quite elusive. In the case of convex surfaces with non-vanishing Gaussian curvature, how much smoothness do we need to obtain optimal pointwise bounds for the Fourier transform of the surface carried measure. The L^2-average decay theorem in Chapter 5 holds under the assumption that the boundary is either convex or $C^{\frac{3}{2}}$. Does the result hold if the boundary is merely Lipschitz? Perhaps even rectifiable boundary is enough? We know that the result fails if we assume that the boundary is merely $d-1$-dimensional, but where is the threshold of roughness? Understanding these issues to a sufficient degree of depth would shed light on many interesting problems described and alluded to in this book, including the celebrated Falconer distance conjecture.

On this note, we thank the reader for his/her patience and temporarily leave the world of exposition for the one of discovery. We hope to be back soon.

Bibliography

[1] C.R. Adams and J.A. Clarkson, *Properties of functions $f(x, y)$ of bounded variation*, Trans. Amer. Math. Soc. **36** (1934), 711–730.

[2] G. Alexits, *Convergence problems of orthogonal series*, Akadémiai Kiadó, Budapest, 1961.

[3] G.I. Arkhipov, V.N. Chubarikov, and A.A. Karatsuba, *Exponent of convergence of the singular integral in the Tarry problem*, (Russian) Dokl. Akad. Nauk SSSR, **248** (1979), 268–272.

[4] G.I. Arkhipov, V.N. Chubarikov, and A.A. Karatsuba, *Trigonometric sums in number theory and analysis*. Translated from the 1987 Russian original. de Gruyter Expositions in Mathematics, 39. Walter de Gruyter GmbH & Co. KG, Berlin, 2004.

[5] V.I. Arnold, A.N. Varchenko, S.M. Gusein-Zade, *Singularities of differentiable maps*, Birkhäuser, 1985.

[6] B. Aubertin, J.J. F. Fournier, *Integrability theorems for trigonometric series*, Studia Math. **107** (1993), 33–59.

[7] K.I. Babenko, *Approximation by trigonometric polynomials in a certain class of periodic functions of several variables*, Dokl. Akad.Nauk SSSR **132** (1960), 982–985 (Russian). – English transl. in Soviet Math. Dokl. **1** (1960), 672–675.

[8] J. Bak, D. McMichael, J. Vance, and S. Wainger, *Fourier transforms of surface area measure on convex surfaces in $\mathbb{R}^3$*, Amer. J. of Math. **111**, (1988), 633–668.

[9] J.-G. Bak and S. Lee, *Estimates for an oscillatory integral operator related to restriction to space curves*, Proc. Amer. Math. Soc. **132** (2004), 1393–1401.

[10] N.K. Bary, *A Treatise on Trigonometric Series, I and II*, MacMillan, New York, 1964.

[11] G. Bateman and A. Erdélyi, *Higher Transcendental Functions, Vol. II*, McGraw Hill Book Company, New York, 1953.

[12] H. Bateman and A. Erdélyi, *Tables of integral transforms, Vol. II*, McGraw Hill Book Company, New York, 1954.

[13] L. Bausov, *On linear methods for the summation of Fourier series*, Mat. Sb. **68** (1965), 313–327 (Russian).

[14] J. Beck, *Irregularities of distribution I*, Acta Math. **159** (1987), 1–49.

[15] J. Beck and W.W.L. Chen, *Irregularities of distribution*, Cambridge University Press, 1987.

[16] J. Beck and W.W.L. Chen, *Note on irregularities of distribution II*, Proc. London Math. Soc. **61** (1990), 102–126.

[17] J. Beck and W.W.L. Chen, *Irregularities of point distribution relative to convex polygons II*, Mathematika **40** (1993), 127–136.

[18] W. Beckner, A. Carbery, S. Semmes, and F. Soria, *A note on restriction of the Fourier transform to spheres*, Bull. London Math. Soc. **21** (1989), 394–398.

[19] E.S. Belinsky, *On asymptotic behavior of integral norms of trigonometric polynomials*, Metric Questions of the Theory of Functions and Mappings, Nauk. Dumka, Kiev **6** (1975), 15–24 (Russian).

[20] E.S. Belinsky, *Application of the Fourier transform to summability of Fourier series*, Sib. Mat. Zh. **XVIII** (1977), 497–511 (Russian). – English transl. in Siberian Math. J. **18** (1977), 353–363.

[21] E. Belinsky, M. Dvejrin, M. Malamud, *Multipliers in L_1 and Estimates for Systems of Differential Operators*, Russian J. Math. Physics **12** (2005), 6–16.

[22] E. Belinsky, E. Liflyand, *Lebesgue constants and integrability of the Fourier transform of radial functions*, Dokl. Acad. Sci. Ukraine, Ser. A **6** (1980), 5–10 (Russian).

[23] E. Belinsky, E. Liflyand, *On asymptotic behavior of Lebesgue constants of radial summability methods*, Constructive Theory of Functions and Theory of Mappings, Nauk. Dumka, Kiev, 1981, 49–70 (Russian).

[24] E. Belinsky, E. Liflyand, R. Trigub, *The Banach algebra A^* and its properties*, J. Fourier Anal. Appl. **3** (1997), 103–129.

[25] E. Belinsky, R.M. Trigub, *A Relation in the Theory of Fourier Series*, Mat. Zametki **15** (1974), 679–682 (Russian). – English transl. in Math. Notes **15** (1974), 405–407.

[26] A. Beurling, *On the spectral synthesis of bounded functions*, Acta math. **81** (1949), 225–238.

[27] R.P. Boas, *Absolute Convergence and Integrability of Trigonometric Series*, J. Rational Mech. Anal. **5** (1956), 621–632.

[28] S.V. Bochkarev, *On a problem of Zygmund*, Izv. Akad. Nauk SSSR, Ser. Mat. **37** (1973), 630–638 (Russian). – English transl. in Math. USSR Izvestija **37** (1973), 629–637.

[29] S.V. Bochkarev, *A Method of Averaging in the Theory of Orthogonal Series and Some Problems in the Theory of Bases*, Trudy Inst. Steklova **146**, 1978 (Russian). – English transl. in Proc. Steklov Inst. Math. **3**, 1980.

[30] S. Bochner, *Summation of multiple Fourier series by spherical means*, Trans. Amer. Math. Soc. **40** (1936), 175–207.

[31] S. Bochner, *Lectures on Fourier Integrals*, Princeton Univ. Press, Princeton, N.J., 1959.

[32] V.A. Borovikov, *Fundamental solution of linear partial differential equations with constant coefficients*, Proc. Moscow Math. Soc. **8** (1959), 199–257 (Russian).

[33] V.A. Borovikov, *Uniform Stationary Phase Method*, The Institution of Electrical Engineers, London, 1994.

[34] V.A. Borovikov, F.J. Mendoza, *The Problem of the Pointwise Fourier Inversion for Piecewise Smooth Functions of Several Variables*, J. Fourier Anal. Appl. **8** (2002), 399–406.

[35] D. Borwein, *Linear functionals connected with strong Cesáro summability*, J. London Math. Soc. **40** (1965), 628–634.

[36] L. Brandolini, *Estimates for Lebesgue Constants in Dimension Two*, Annali di Matematica Pura e Applicata (IV) **CLVI** (1990), 231–242.

[37] L. Brandolini, *Fourier transform of characteristic functions and Lebesgue constants for multiple Fourier series*, Colloquium Math. **LXV** (1993), 51–59.

[38] L. Brandolini, L. Colzani, *Localization and convergence of eigenfunction expansions*, Journ. Fourier Anal. Appl. **5** (1999), 431–447.

[39] L. Brandolini, L. Colzani, *Decay of Fourier Transforms and Summability of Eigenfunction Expansions*, Ann. Scuola Norm. Sup. Pisa Cl. Sci (4) 29 (2000), 611–638.

[40] L. Brandolini, L. Colzani and G. Travaglini, *Average decay of Fourier transforms and integer points in polyhedra*, Ark. f. Mat. **35** (1997), 253–275.

[41] L. Brandolini, M. Rigoli and G. Travaglini, *Average decay of Fourier transforms and geometry of convex sets*, Revista Mat. Iberoam. **14** (1998), 519–560.

[42] L. Brandolini and G. Travaglini, *Pointwise convergence of Fejer type means*, Tohoku Math. J. **49** (1997), 323–336.

[43] L. Brandolini, A. Greenleaf and G. Travaglini, $L^p - L^{p'}$ *estimates for overdetermined Radon transforms*, Trans. Amer. Math. Soc., to appear.

[44] L. Brandolini, L. Colzani, A. Iosevich, A.N. Podkorytov, and G. Travaglini, *Geometry of the Gauss map and lattice points in convex domain*, Mathematika **48** (2001), 107–117.

[45] L. Brandolini, S. Hofmann and A. Iosevich, *Sharp rate of average decay of the Fourier transform of a bounded set*, Geom. Funct. Anal. **13** (2003), 671–680.

[46] J. Bruna, A. Nagel and S. Wainger, *Convex hypersurfaces and Fourier transforms*, Ann. Math. **127** (1988), 333–365.

[47] M. Buntinas and N. Tanović-Miller, *New integrability and* L^1*-convergence classes for even trigonometric series II*, Approximation Theory (J. Szaba-

dos and K. Tandori, eds.), Colloq. Math. Soc. János Bolyai, North-Holland, Amsterdam **58** (1991), 103–125.

[48] P.L. Butzer, R.J. Nessel and W. Trebels, *On radial M_p^q Fourier multipliers*, Math. Struct., Comput. Math., Math. Modelling, Sofia, 1975, 187–193.

[49] A. Carbery, *A note on the "hyperbolic" Bochner–Riesz means*, Proc. Amer. Math. Soc. **92** (1984), 397–400.

[50] A. Carbery, M. Christ and J. Wright, *Multidimensional van der Corput and sublevel set estimates*, J. Amer. Math. Soc. **12** (1999), 981–1015.

[51] L. Carleson and P. Sjölin, *Oscillatory integrals and a multiplier problem for the disc*, Collection of articles honoring the completion by Antoni Zygmund of 50 years of scientific activity, III. Studia Math. **44** (1972), 287–299.

[52] W.W.L. Chen, *On irregularities of distribution III*, J. Austr. Math. Soc. **60** (1996), 228–244.

[53] M. Christ, *On the restriction of the Fourier transform to curves: endpoint results and the degenerate case*, Trans. Amer. Math. Soc. **287**, (1985), 223–238.

[54] R. Coifman and G. Weiss, *Extensions of Hardy spaces and their use in analysis*, Bull. Amer. Math. Soc. **83** (1977), 569–645.

[55] Y. Colin de Verdière, *Nombre de points entiers dans une famille homothétique de domains de R*, Ann. Sci. Ecole Norm. Sup. **10** (1977), 559–575.

[56] L. Colzani, *Fourier transform of distributions in Hardy Spaces*, Boll. U.M.I. **1** (1982), 403–410.

[57] E.T. Copson, *Asymptotic Expansions*, Cambridge Univ. Press, Cambridge, 1965.

[58] A. Cordoba, *La formule sommatoire de Poisson*, C. R. Acad. Paris, Ser. I **306** (1988), 373–376.

[59] J. Cossar, *A theorem on Cesàro summability*, J. London Math. Soc. **16** (1941), 56–68.

[60] H. Davenport, *Note on irregularities of distribution*, Mathematika **1** (1954), 73–79.

[61] K.M. Davis, Y.-C. Chang, *Lectures on Bochner–Riesz means*, London Math. Soc. Lecture Note Series 114, Cambridge Univ. Press, Cambridge, 1987.

[62] S.W. Drury, *Restriction of Fourier transforms to curves*, Ann. Inst. Fourier, **35** (1985), 117–123.

[63] J. Duoandikoetxea, *Fourier Analysis*, Graduate Studies in Mathematics, 29. Amer. Math. Soc., Providence, RI, 2001.

[64] A. El-Kohen, *On the hyperbolic Riesz means*, Proc. Amer. Math. Soc. **89** (1983), 113–116.

[65] R.E. Edwards, *Criteria for Fourier transforms*, J. Australian Math. Soc. **7** (1967), 239–246.

[66] A. Erdélyi, *Asymptotic expansions*, Dover, New York, 1956.

[67] D. Faifman, *A characterization of Fourier transform by Poisson summation formula*, C. R. Acad. Paris, Ser. I **348** (2010), 407–410.

[68] K.J. Falconer, *On the Hausdorff dimensions of distance sets*, Mathematika **32**, (1986), 206–212.

[69] D. Fan, Y. Pan, *On the Fourier transform of measures carried by submanifolds of finite type*, J. d'Analyse Math. **71** (1997), 135–147.

[70] M.V. Fedoryuk, *The stationary phase method and pseudodifferential operators*, Uspekhi Mat. Nauk **26** (1971), 67–112 (Russian). – English transl. in Soviet Math. Surveys **26** (1971), 65–115.

[71] C. Fefferman, *Inequalities for strongly singular convolution operators*, Acta Math. **124** (1970), 9–36.

[72] H. Feichtinger, *Wiener amalgams over Euclidean spaces and some of their applications*, Function spaces (Edwardsville, IL, 1990), Lect. Notes Pure Appl. Math., Dekker, New York **136**, 123–137 (1992).

[73] T.M. Flett, *Some elementary inequalities for integrals with applications to Fourier transforms*, Proc. London Math. Soc. (3), **29** (1974), 538–556.

[74] G.A. Fomin, *A Class of Trigonometric Series*, Matem. Zametki **23** (1978), 213–222 (Russian). – English transl. in Math. Notes **23** (1978), 117–123.

[75] J.J.F. Fournier, J. Stewart, *Amalgams of L^p and ℓ^q*, Bull. Amer. Math. Soc. **13** (1985), 1–21.

[76] S. Fridli, *Integrability and L^1-convergence of trigonometric and Walsh series*, Annales Univ. Sci. Budapest, Sect. Comp. **16** (1996), 149–172.

[77] S. Fridli, *Hardy spaces generated by an integrability condition*, J. Approx. Theory **113** (2001), 91–109.

[78] M. Ganzburg, E. Liflyand, *Estimates of Best Approximation and Fourier Transforms in Integral Metrics*, J. Approx. Theory **83** (1995), 347–370.

[79] G. Gasper and W. Trebels, *A characterization of localized Bessel potential spaces and applications to Jacobi and Hankel multipliers*, Studia Math. **65** (1979), 243–278.

[80] G. Gasper and W. Trebels, *Multiplier criteria of Hörmander type for Fourier series and applications to Jacobi series and Hankel transforms*, Math. Ann. **242** (1979), 225–240.

[81] I.M. Gelfand, M.I. Graev and N.Ya. Vilenkin, *Generalized functions, Vol. 5: Integral geometry and problems of representation theory*, Acad. Press, New York, 1966.

[82] D.V. Giang and F. Móricz, *On the integrability of trigonometric series*, Anal. Math. **18** (1992), 15–23.

[83] D.V. Giang and F. Móricz, *Lebesgue integrability of double Fourier transforms*, Acta Sci. Math. (Szeged) **58** (1993), 299–328.

[84] D.V. Giang and F. Móricz, *Multipliers of double Fourier transforms and Fourier series on L^1*, Acta Sci. Math. (Szeged) **58** (1993), 329–348.

[85] D.V. Giang and F. Móricz, *Multipliers of Fourier transforms and series on L^1*, Arch. Math. **62** (1994), 230–238.

[86] E. Giusti, *Minimal surfaces and functions of bounded variation*, Birkhäuser, Boston, 1984.

[87] M.L. Goldman, *Estimates for Multiple Fourier Transforms of Radially Symmetric Monotone Functions*, Sib. Mat. Zh. **18** (1977), 549–569 (Russian). – English transl. in Sib. Math. J. **18** (1977), 391–406.

[88] L. Grafakos, Classical Fourier Analysis, 2nd ed., Springer, 2008.

[89] L. Grafakos, Modern Fourier Analysis, 2nd ed., Springer, 2009.

[90] A. Greenleaf, A. Seeger and S. Wainger, *On X-ray transforms for rigid line complexes and integrals over curves in R^4*, Proc. Amer. Math. Soc. **127** (1999), 3533–3545.

[91] G.H. Hardy, *Oscillating Dirichlet integrals*, Quart. J. Pure Appl. Math. **44** (1913), 1–40, 242–263.

[92] C. Heil, *An Introduction to Weighted Wiener Amalgams*, Wavelets and their Applications (Chennai, 2002), M. Krishna, R. Radha and S. Thangavelu, eds., Allied Publishers, New Delhi, 183–216 (2003).

[93] C.S. Herz, *On the number of lattice points in convex set*, Amer. J. Math. **84** (1962), 126–133.

[94] C.S. Herz, *Fourier transforms related to convex sets*, Ann. Math. (2)**75** (1962), 81–92.

[95] C.S. Herz, *Lipschitz spaces and Bernstein's theorem on absolutely convergent Fourier transforms*, J. Math. Mech. **18** (1968/69), 283–323.

[96] E. Hlawka, *Über Integrale auf konvexen Körpern. I & II*, Monatsh. Math. **54** (1950), 1–36, 81–99.

[97] E.W. Hobson, *Theory of Functions of a Real Variable*, 3rd edition, Vol. I, Cambridge, 1927.

[98] S. Hofmann and A. Iosevich, *Falconer conjecture in the plane for random metrics*, Proc. Amer. Math. Soc. **133** (2005), 133–143.

[99] F. Holland, *Harmonic analysis on amalgams of L^p and l^q*, J. London Math. Soc. (2)**10** (1975), 295–305.

[100] L. Hörmander, *Oscillatory integrals and multipliers on FL^p*, Ark. Mat. **11** (1973), 1–11.

[101] L. Hörmander, *The analysis of linear partial differential operators. I. Distribution theory and Fourier analysis.* Reprint of the second (1990) edition. Classics in Mathematics. Springer-Verlag, Berlin, 2003.

[102] M.N. Huxley, *Area, lattice points, and exponential sums*, Clarendon Press, Oxford, 1996.

[103] I.A. Ikromov, *On an estimate for the Fourier transform of the indicator of nonconvex domains*, Funktsional. Anal. i Ego Prilozhen. **29** (1995), 16–24 (Russian). – English transl. in Funct. Anal. Appl. **29** (1995), 161–167.

[104] I.A. Ikromov and D. Müller, *Uniform Estimates for the Fourier Transform of Surface Carried Measures in $\mathbb{R}^3$ and an Application to Fourier Restriction*, J. Fourier Anal. Appl. **17** (2011), 1292–1332.

[105] V.A. Ilyin and S.A. Alimov, *Conditions for the convergence of spectral decompositions that correspond to self-adjoint extensions of elliptic operators, I*, Diff. Eq. **7** (1971), 516–543.

[106] A. Iosevich, *Generalized Diophantine conditions and the distribution of lattice points in convex planar domains*, J. Number Theory **90** (2001), 19–30.

[107] A. Iosevich, *Lattice points and generalized diophantine conditions*, Journal of Number Theory, **90**, no. 1 (2001), 19–30.

[108] A. Iosevich and I. Laba, *K-distance sets and the Falconer conjecture*, Integers **5** (2005), 11 pp.

[109] A. Iosevich and E. Sawyer, *Averages over surfaces*, Adv. in Math. **132**, (1997), 46–119.

[110] A. Iosevich and E. Sawyer, *Three problems motivated by the average decay of the Fourier transform*, Harmonic analysis at Mount Holyoke (South Hadley, MA, 2001), 205–215, Contemp. Math., 320, Amer. Math. Soc., Providence, RI, 2003.

[111] S.I. Izumi and T. Tsuchikura, *Absolute convergence of trigonometric expansions*, Tôhoku Math. J. **7** (1955), 243–251.

[112] V. Jarnik, *Über die Gitterpunkte auf konvexen Kurven*, Math. Zeitschrift **24** (1925), 500–518.

[113] J.-P. Kahane, *Séries de Fourier absolument convergentes*, Springer-Verlag, 1970.

[114] D.G. Kendall, *On the number of lattice points in a random oval*, Quart. J. Math Oxford Ser. **19** (1948), 1–26.

[115] M. Kolountzakis and T. Wolff, *On the Steinhaus tiling problem*, Mathematika **46** (2002), 253–280.

[116] E. Krätzel, *Lattice points*, Kluwer Academic Publisher, 1988.

[117] E. Landau, *Über die Gitterpunkte in einem Kreise* (II), Göttinger Nachrichten, 1915, 161–171.

[118] E. Liflyand, *On some conditions of integrability of the Fourier transform*, Ukr. Math. J. **32** (1980), 110–118 (Russian).

[119] E. Liflyand, *On asymptotics of Fourier transform for functions of certain classes*, Anal. Math. **19** (1993), 151–168.

[120] E. Liflyand, *Fourier Transforms of Radial Functions*, Integral Transforms and Special Functions **4** (1996), 279–300.

[121] E. Liflyand, *On the Bochner–Riesz means of critical order*, Proc. Amer. Math. Soc. **125** (1997), 1443–1450.

[122] E. Liflyand, *A Decrease in Growth of Lebesgue Constants*, J. Math. Anal. Appl. **212** (1997), 20–29.

[123] E. Liflyand, *A family of function spaces and multipliers*, Israel Math. Conf. Proc. **13** (1999), 141–149.

[124] E. Liflyand, *On Bausov–Telyakovskii type conditions*, Georgian Math. J. **47** (2000), 753–764.

[125] E. Liflyand, *Lebesgue constants of multiple Fourier series*, Online J. Anal. Combin. **1** (2006), 112 pp.

[126] E. Liflyand, *A Counterexample on Embedding of Spaces*, Georgian Math. J., **16** (2009), 757–760.

[127] E. Liflyand, *Fourier transform versus Hilbert transform*, Ukr. Math. Bull. **9** (2012), 209–218. Also published in J. Math. Sciences **187** (2012), 49–56.

[128] E. Liflyand, *Fourier transforms on an amalgam type space*, Monatsh. Math. **172** (2013), 345–355.

[129] E. Liflyand, *Interaction between the Fourier transform and the Hilbert transform*, Acta Comm. Uni. Tartu. Math. **18** (2014).

[130] E. Liflyand, A. Nakhman, *On linear means of multiple Fourier integrals defined by special domains*, Rocky Mountain J. Math. **32** (2002), 969–980.

[131] E. Liflyand, A.G. Ramm, A.I. Zaslavsky, *Estimates from below for Lebesgue constants*, J. Fourier Anal. Appl. **2** (1996), 287–301.

[132] E. Liflyand, M. Skopina, *Square linear means with hyperbolic factors*, Analysis **18** (1998), 333–343.

[133] E. Liflyand, S. Samko and R. Trigub, *The Wiener Algebra of Absolutely Convergent Fourier Integrals: an overview*, Anal. Math. Physics **2** (2012), 1–68.

[134] E. Liflyand, U. Stadtmüller and R. Trigub, *An interplay of multidimensional variations in Fourier Analysis*, J. Fourier Anal. Appl. **17** (2011), 226–239.

[135] E. Liflyand and U. Stadtmüller, *A Multidimensional Euler–Maclaurin Formula and an Application*, Contemporary Mathematics **591** (2013), 183–194.

[136] E. Liflyand, W. Trebels, *On Asymptotics for a Class of Radial Fourier Transforms*, Zeitschrift für Analysis und ihre Anwendungen (J. Anal. Appl.) **17** (1998), 103–114.

[137] W. Littman, *Fourier transforms of surface carried measures, and differentiability of surface averages*, Bull. Amer. Math. Soc. **69** (1963), 766–770.

[138] J. Löfstöm, *Some theorems on interpolation spaces with application to approximation on L_p*, Math. Ann. **172** (1967), 176–196.

[139] E. Lucacs, *Characteristic functions, 2nd ed.*, Charles Griffin & Co. Ltd., London, 1970.

[140] B. Makarov and A. Podkorytov, *Real Analysis: Measures, Integrals and Applications*, Springer, 2013.

[141] B.P. Marshall, Decay rates of Fourier transforms of curves, *Trans. Amer. Math. Soc.* **310** (1988), 115–126.

[142] P. Mattila, *Spherical averages of Fourier transforms of measures with finite energy; dimension of intersections and distance sets*, Mathematika **34** (1987), 207–228.

[143] V.G. Maz'ya and T.O. Shaposhnikova, *Theory of Sobolev Multipliers With Applications to Differential and Integral Operators*, Springer-Verlag Berlin Heidelberg, 2009.

[144] J. Milnor, *Morse theory*, Princeton Univ. Press, Princeton, 1963.

[145] A. Miyachi, *On some Fourier multipliers for $H^p(R^n)$*, J. Fac. Sci. Univ. Tokyo Sec. IA, **27** (1980), 157–179.

[146] S. Mizohata, *The theory of partial differential equations*, Cambridge Univ. Press, 1973.

[147] G. Mockenhaupt, *Bounds in Lebesgue spaces of oscillatory integral operators.* Habilitationsschrift, Universität Siegen (1996).

[148] H.L. Montgomery, *Ten lessons on the interface between analytic number theory and harmonic analysis*, CBMS Regional Conference Series in Mathematics, **84**, American Mathematical Society, 1994.

[149] F. Móricz, *On the integrability and L^1-convergence of complex trigonometric series*, Proc. Amer. Math. Soc. **113** (1991), 53–64.

[150] D. Müller and A. Seeger, *Regularity properties of wave propagation on conic manifolds and applications to spectral multipliers*, Adv. Math., **161** (2001), 41–130.

[151] A. Nagel, A. Seeger and S. Wainger, *Averages over convex hypersurfaces*, Amer. J. Math. **115** (1993), 903–927.

[152] F.L. Nazarov and A.N. Podkorytov, *On the behaviour of the Lebesgue constants for two-dimensional Fourier sums over polygons*, St.-Petersburg Math. J. **7** (1996), 663–680.

[153] D. Oberlin, *Oscillatory integrals with polynomial phase*, Math. Scand. **69** (1991), 45–56.

[154] A.N. Podkorytov, *Linear means of spherical Fourier sums*, Operator Theory and Function Theory, ed. M.Z. Solomyak, Leningrad Univ., Leningrad **1** (1983), 171–177 (Russian).

[155] A.N. Podkorytov, *The asymptotics of a Fourier transform on a convex curve*, Vestnik Leningrad Univ. Matem. **24** (1991), 50–57 (Russian). – English transl. in Vestnik Leningrad Univ. Math. **24** (1991), 57–65.

[156] A.G. Ramm, A.I. Katsevich, *The Radon transform and local tomography*, CRC Press, Boca Raton, 1996.

[157] A.G. Ramm and A.I. Zaslavsky, *Asymptotic behavior of the Fourier transform of piecewise-smooth functions*, C. R. Acad. Sci. Paris **316** (1993), 541–546.

[158] A.G. Ramm and A.I. Zaslavsky, *Singularities of the Radon transform*, Bull. Amer. Math. Soc. **25** (1993), 109–115.

[159] A.G. Ramm and A.I. Zaslavsky, *Reconstructing singularities of a function from its Radon transform*, Math. Comp. Modelling **18** (1993), 109–138.

[160] B. Randol, *A lattice-point problem II*, Trans. Amer. Math. Soc. **125** (1966) 101–113.

[161] B. Randol, *On the Fourier transform of the indicator function of a planar set*, Trans. Amer. Math. Soc. **139** (1969), 271–278.

[162] B. Randol, *On the asymptotic behavior of the Fourier transform of the indicator function of a convex set*, Trans. Amer. Math. Soc. **139** (1969), 279–285.

[163] F. Ricci and G. Travaglini, *Convex curves, Radon transforms and convolution operators defined by singular measures*, Proc. Amer. Math. Soc. **129** (2001), 1739–1744.

[164] P.G. Rooney, *On the representation of sequences as Fourier coefficients*, Proc. Amer. Math. Soc. **11** (1960), 762–768.

[165] R. Ryan, *Fourier transforms of certain classes of integrable functions*, Trans. Amer. Math. Soc. **105** (1962), 102–111.

[166] R. Salem, *Essais sur les séries trigonometriques*, Actual. Sci. et Industr. **862**, Paris, 1940.

[167] S.G. Samko, A.A. Kilbas, O.I. Marichev, *Fractional Integrals and Derivatives. Theory and Applications*, Gordon& Breach Sci. Publ., New York, 1992.

[168] S.G. Samko, G.S. Kostetskaya, *Absolute integrability of Fourier integrals*, Vestnik RUDN (Russian). Univ. of Peoples Friendship, Ser. Matem. **1** (1994), 138–168.

[169] H. Schulz, *Convex Hypersurfaces of Finite Type and the Asymptotics of Their Fourier Transforms*, Indiana Univ. Math. J. **40** (1991), 1267–1275.

[170] C. Schütt, *The convex floating body and polyhedral approximation*, Israel J. Math. **73** (1991), 65–77.

[171] A. Seeger and S. Ziesler, *Riesz means associated with convex domains in the plane.* Math. Z. **236** (2001), 643–676.

[172] G.E. Shilov, *On the Fourier coefficients of a class of continuous functions*, Doklady AN SSSR **35** (1942), 3–7 (Russian).

[173] S. Sidon, *Reihentheoretische Sätze und ihre Anwendungen in der Theorie der Fourierischen Reihen*, Math. Zeitschrift **10** (1921), 121–127.

[174] P. Sjölin, *Estimates of spherical averages of Fourier transforms and dimensions of sets*, Mathematika **40** (1993), 322–330.

[175] M.M. Skriganov, *Ergodic theory on $SL(n)$, Diophantine approximations and anomalies in the lattice point problem*, Invent. Math. **132** (1998), 1–72.

[176] W.T. Sledd and D.A. Stegenga, *An H^1 multiplier theorem*, Ark. Mat. **19** (1981), 265–270.

[177] C.D. Sogge, *Fourier Integrals in Classical Analysis*, Cambridge Univ. Press, 1993.

[178] E.M. Stein, *On certain exponential sums arising in multiple Fourier series*, Ann. Math. **73** (1961), 87–109.

[179] E.M. Stein, *Singular Integrals and Differentiability Properties of Functions*, Princeton Univ. Press, Princeton, N.J., 1970.

[180] E.M. Stein, *Harmonic Analysis: Real-Variable Methods, Orthogonality, and Oscillatory Integrals*, Princeton Univ. Press, Princeton, N.J., 1993.

[181] E.M. Stein and G. Weiss, *Introduction to Fourier Analysis on Euclidean Spaces*, Princeton Univ. Press, Princeton, N.J., 1971.

[182] E.M. Stein, R. Shakarchi, *Complex analysis.* Princeton Lectures in Analysis, II. Princeton University Press, Princeton, N.J., 2003.

[183] R. Strichartz, *Harmonic analysis as spectral theory of laplacians*, J. of Func. Anal. **87** (1989), 51–148.

[184] R. Strichartz, *Fourier asymptotics of fractal measures*, J. Funct. Anal. **89**, (1990), 154–187.

[185] I. Svensson, *Estimates for the Fourier transform of the characteristic function of a convex set*, Ark. för Mat. **9** (1971), 11–22.

[186] N. Tanović-Miller, *On a paper of Bojanić and Stanojević*, Rendiconti del Circolo Matematico di Palermo, Serie II **XXXIV** (1985), 310–324.

[187] M. Tarnopolska-Weiss, *On the number of lattice points in a compact n-dimensional polyhedron*, Proc. Amer. Math. Soc. **74** (1979), 124–127.

[188] S.A. Telyakovskii, *Integrability conditions for trigonometric series and their applications to the study of linear summation methods of Fourier series*, Izv. Akad. Nauk SSSR, Ser. Matem. **28** (1964), 1209–1236 (Russian).

[189] S.A. Telyakovskii, *An asymptotic estimate for the integral of the absolute value of a function, given by the sine series*, Sibirsk. Mat. Zh. **8** (1967), 1416–1422 (Russian).

[190] S.A. Telyakovskii, *Estimates of trigonometric series and polynomials in connection with problems of the theory of approximation of functions; MS thesis*, Matem. Zametki **1** (1967), 611–623 (Russian).

[191] S.A. Telyakovskii, *An estimate, useful in problems of approximation theory, of the norm of a function by means of its Fourier coefficients*, Trudy Mat. Inst. Steklova **109** (1971), 65–97 (Russian). – English transl. in Proc. Steklov Math. Inst. **109** (1971), 73–109.

[192] S.A. Telyakovskii, *Concerning a sufficient condition of Sidon for the integrability of trigonometric series*, Mat. Zametki **14** (1973), 317–328 (Russian). – English transl. in Math. Notes **14** (1973), 742–748.

[193] E.C. Titchmarsh, *Introduction to the theory of Fourier integrals*, Oxford, 1937.

[194] W. Trebels, *Multipliers for (C,α)-Bounded Fourier Expansions in Banach Spaces and Approximation Theory*, Lect. Notes Math. 329, Springer, Berin, 1973.

[195] W. Trebels, *Some Fourier multiplier criteria and the spherical Bochner–Riesz kernel*, Rev. Roum. Math. Pures et Appl. **20** (1975), 1173–1185.

[196] R.M. Trigub, *A relation between summability and absolute convergence of Fourier series and transforms*, Dokl. Akad. Nauk SSSR **217** (1974), 34–37 (Russian). – English transl. in Soviet Math. Dokl. **15** (1974), 1015–1019.

[197] R.M. Trigub, *Integrability of Fourier transform of boundedly supported function*, Theory of Functions, Functional Analysis and their Applications **23** (1974), 124–131 (Russian).

[198] R.M. Trigub, *On integral norms of polynomials*, Matem. Sbornik **101(143)** (1976), 315–333(Russian). – English transl. in Math. USSR Sbornik **30** (1976), 279–295.

[199] R.M. Trigub, *Integrability and asymptotic behavior of the Fourier transform of a radial function*, Metric Questions of Theory of Functions and Mappings, Kiev, Nauk. dumka, 1977, 142–163 (Russian).

[200] R.M. Trigub, *Linear methods of summability of simple and multiple Fourier series and their approximate properties*, Proc. Intern. Conf. Nauka, Moscow, 1977, 383–390 (Russian).

[201] R.M. Trigub, *Multipliers of Fourier series and approximation of functions by polynomials in spaces C and L*, Dokl. Akad. Nauk SSSR **306** (1989), 292–296 (Russian). – English transl. in Soviet Math. Dokl. **39** (1989), 494–498.

[202] R.M. Trigub, *A Generalization of the Euler–Maclaurin formula*, Mat. Zametki **61** (1997), 312–316 (Russian). – English transl. in Math. Notes **61** (1997), 253–257.

[203] R.M.Trigub and E.S. Belinsky, *Fourier Analysis and Approximation of Functions*, Kluwer, 2004.

[204] A.N. Varchenko, *Number of lattice points in families of homothetic domains*, Funk. Anal. i Ego Pril. **17** (1983), 1–6 (Russian). – English transl. in Funct. Anal. Appl. **17** (1983), 79–83.

[205] R. Wheeden and A. Zygmund, *Measure and Integral*, Pure and Applied Mathematics, (1977).

[206] D.V. Widder, *The Laplace Transform*, Princeton, 1946.

[207] N. Wiener, *On the representation of functions by trigonometric integrals*, Math. Z. **24** (1926), 575–616.

[208] N. Wiener, *The Fourier integral and certain of its applications*, Dover Publ., Inc., New York, 1932.

[209] T. Wolff, *Math* 191 *lecture notes*, Harmonic analysis lecture notes; CalTech (2000).

[210] A.A. Yudin and V.A. Yudin, *Polygonal Dirichlet kernels and growth of Lebesgue constants.* Mat. Zametki **37** (1985), 220–236 (Russian). – English transl. in Math. Notes **37** (1985), 124–135.

[211] V.A. Yudin, *Lower bound for Lebesgue constants*, Mat. Zametki **25** (1979), 119–122 (Russian). – English transl. in Math. Notes **25** (1979), 63–65.

[212] W.H. Young and G.C. Young, *On the discontinuities of monotone functions of several variables*, Proc. London Math. Soc., (2) **22** (1923), 124–142.

[213] A.I. Zaslavsky, *Dual varieties and Legendre transforms*, J. Math. Anal. Appl. **183** (1994), 670–684.

[214] A.I. Zaslavsky, *Multidimensional analogue of the Erdélyi lemma and the Radon transform*, Tomography, impedance imaging and integral geometry (South Hadley, MA, 1993), Lectures Appl. Math. 30, Amer. Math. Soc., Providence, RI, 1994, 259–278.

[215] A. Zygmund, *Trigonometric series*, Cambridge Univ. Press, New York, 1959.

Index

MIX
Papier aus verantwortungsvollen Quellen
Paper from responsible sources
FSC® C105338

If you have any concerns about our products,
you can contact us on
ProductSafety@springernature.com

In case Publisher is established outside the EU,
the EU authorized representative is:
Springer Nature Customer Service Center GmbH
Europaplatz 3, 69115 Heidelberg, Germany

Printed by Libri Plureos GmbH
in Hamburg, Germany